THE ENERGY-ECONOMY LINK

THE ENERGY-ECONOMY LINK

New Strategies for the Asia-Pacific Region

Edited by
WILLIAM E. JAMES

PRAEGER

New York
Westport, Connecticut
London

Library of Congress Cataloging-in-Publication Data

The Energy-economy link : new strategies for the Asia–Pacific region / edited by William E. James.
p. cm.
Papers from a seminar held August 3–7, 1987, in Bangkok, Thailand.
Bibliography: p.
Includes index.
ISBN 0-275-93104-8 (alk. paper)
1. Energy development—Asia—Planning—Congresses. 2. Energy development—Pacific Area—Planning—Congresses. 3. Asia—Industries—Energy consumption—Congresses. 4. Pacific Area—Industries—Energy consumption—Congresses. 5. Energy policy—Asia—Congresses. 6. Energy policy—Pacific Area—Congresses. 7. Asia—Economic policy—Congresses. 8. Pacific Area—Economic policy—Congresses. I. James, William E.
HD9502.A782E535 1989
333.79'095—dc19 88-38828

Library of Congress Catalog Card Number: 88-38828
ISBN: 0-275-93104-8

First published in 1989

Praeger Publishers, One Madison Avenue, New York, NY 10010
A division of Greenwood Press, Inc.

Printed in the United States of America

The paper used in this book complies with the Permanent Paper Standard issued by the National Information Standards Organization (Z39.48–1984).

10 9 8 7 6 5 4 3 2 1

Contents

Foreword

The seminar "Energy Sector Performance and Economic Development in Asia" was held August 3–7, 1987, in Bangkok, Thailand. Organizations involved in the conduct of the program included the Asian Productivity Organization, the East-West Center's Resource Systems Institute, and the Thailand Management, Development, and Productivity Center.

The main objective of the seminar was to enable participants to identify the major linkages between the energy sector and economic development, to assess the role of the energy sector in future development, and to discuss energy policy options that will optimize economic growth.

Through this publication the organizers would like to share the success of the seminar with colleagues. We hope that it will prove useful to those concerned with the energy-macroeconomy link of growth and development. The following text includes papers presented by core faculty and invited experts, as well as selected country papers presented by participants.

We wish to express our appreciation to Dr. William E. James, seminar co-chairman, who oversaw the preparation of the book; to the authors, for their continued cooperation while the text was being revised; and to David Puhlick, Resource Systems Institute editor, for putting the manuscript into its final form.

On behalf of our organizations, we wish to express our appreciation to the Thailand Management, Development, and Productivity Center for its role in planning and organizing the seminar.

Nagao Yoshida
Secretary-General
Asian Productivity Organization

Seiji Naya
Director
Resource Systems Institute
East-West Center

Introduction

DR. R. K. PACHAURI

Energy developments in Asia have followed a varying path since the mid–1970s. The first oil price shock of 1973–1974 resulted in a range of immediate measures, in some cases bordering on panic reactions, but the second oil price shock of 1979–1980 was, in general, tackled with a greater degree of confidence on the part of governments in the region, even though in several respects the effects of this price shock were far more severe than those of the first. Steps taken following the decline in oil prices in 1986 obviously have not run their full course, but a broad analysis of them can be attempted even with the limited experience that we have with post–1986 policies and their responses.

This volume carries in-depth studies on several countries of the region, and therefore the general review presented in this introductory chapter may be at variance with specific country analyses presented in the subsequent chapters. However, some generalizations and broad conclusions for Asia as a whole are relevant in assessing the overall post–1986 developments within this continent.

One important message that is available from a review of Asia as a whole is that directions and policies that evolved in the mid–1970s have not been altered in any significant manner with the downturn in global oil prices in 1986. In fact, a study undertaken by the Asian Development Bank indicates that the major concerns of governments in the region still include energy self-reliance, energy security, and oil substitution. These objectives do not appear to differ significantly from those that may have been enunciated in 1974. Nor has there been any major change in the strategies for economic development that were in evidence between 1973–1974 and 1986. Consequently, the overall effort to develop indigenous supplies of energy or to diversify sources and forms of imported energy has not changed since 1986.

The share of oil in total primary commercial energy supply has declined in Bangladesh, Pakistan, Sri Lanka, the Philippines, Thailand, Taiwan, and the Republic of Korea. This has occurred largely because of an increase in the share of natural gas in Bangladesh, Pakistan, and Thailand, and a slight increase in hydroelectric energy in Sri Lanka and the Philippines. Taiwan and the Republic of Korea continue with the tendency toward an increase in the share of nuclear energy. The Philippines, Thailand, and Taiwan also show an increase in the share of coal imports as part of long-run policy initiatives. On the other hand, India and Nepal continue to exhibit increases in the share of oil as part of the total energy supply.

Major changes have also taken place on the demand side that result from structural shifts in the economy as well as changes in energy use intensity. The trend toward reduction in the energy/GDP ratio has not been altered by a decline in international oil prices. However, structural changes have occurred, leading, in several cases, to an increase in energy use intensity. This fact also underlines the reality that the path of economic development adopted by the countries of the region perhaps has not been affected by changes in international oil prices. In other words, the process of industrialization, favoring an expansion of heavy and energy-intensive industries in many countries, has not changed significantly with the changes in international oil prices. Among different consuming sectors, industry continues to be the largest user in relation to the transportation, residential, and commercial sectors. However, the power sector, which is essentially a converter rather than a consumer of energy, accounts for the largest share of primary energy consumption, including that of the industrial sector. Current trends are toward an expansion of the power sector and its share in total energy consumption in relation to other sectors of the economy. Investment plans already in hand for expansion of the power sector have not been altered significantly with the drop in oil prices during 1986; therefore it may be too early to arrive at any conclusions on this issue.

In essence, one could summarize the Asian experience since 1986 by stating that the governments of the region have generally not made significant policy changes, and their priorities are still determined by the effects of the first and second oil price shocks. There are, however, very interesting differences in strategy between the different countries, and very striking differences in emphasis with regard to fuel mix, the pattern of industrialization, and development of energy-consuming sectors. These differences are discussed in some depth in subsequent chapters, but they do not, in general, detract from the reality that governments have in general been cautious in altering paths charted by them in the mid- and late 1970s.

THE ENERGY-ECONOMY LINK

PART I
Energy and Economic Development in the Region

ONE

The International Petroleum Market: Changing Structure and the Future Outlook

FEREIDUN FESHARAKI

INTRODUCTION

Much has happened in the oil market since the Tehran Conferences in 1971, when the price of oil increased for the first time in more than 15 years. The shift in control of oil resources from the oil companies to the producing governments, the oil price shocks of 1973–1974 and 1979–1980, and the massive slide in oil demand in the early 1980s have resulted in structural changes in the oil market that have fundamentally altered the relationship between producers and consumers. These structural changes have not yet ended. While the oil industry and the governments (on both sides) remain preoccupied with short-term supply, and demand and price questions, the underlying changes slowly continue to alter the structure of the oil market.

This chapter attempts to analyze the underlying factors likely to shape the oil market and the oil political economy over the next two decades. The author, who has long been engaged in statistical analysis and oil market forecasting, has deliberately decided to avoid excessive reliance on statistical analysis. Rather, this chapter represents his subjective perception, with a view toward the long-term factors affecting our lives and the lives of our children.

WHY OIL FORECASTS GO WRONG

Oil marketing forecasting, despite its poor past performance, remains an integral part of energy policy. Producer (net exporter) and consumer (net importer) governments, the oil industry, and all other sectors of the world economy that depend on energy rely on oil forecasts to make policy and investment decisions.

Today's policy and investment decisions shape the future, yet today's forecasts are based on yesterday's information—and poor information it is.

Oil market forecasting is difficult because the information is imperfect, forecasting techniques are far from adequate, and subjective biases are often introduced into the assumptions.[1] It is thus no great wonder that oil forecasts are often wrong. Let us take a closer look at the factors affecting our outlook on oil.

First is *imperfect information*—we do not have current data. Our data for most of the industrial world are at best a year old, and for the developing world, two to three years old. At the same time, many governments and oil companies treat data on oil as a security concern and are reluctant to make them available in a useful form or timely manner.

Second is *inadequate forecasting techniques*—econometrics applied to oil market forecasting with a poor information base cannot be seriously relied upon. More scientific techniques, such as end-use analysis, require far more elaborate data than are available.

Perceptions

Projections of the future of the oil market are affected more by the perceptions of the analysts than by the actual underlying changes in demand and supply. Clearly, perceptions differ, as do assumptions about underlying changes in the market, leading to differing projections for the future. Perceptions affect oil market forecasts in much the same way they affect the stock market. That is, short-term considerations are projected long into the future, further affecting immediate decisions. We can list a series of perception changes that have occurred in the oil market since the mid–1970s:

- Pre–1973—Perceptions of indefinite supply of oil, rising demand, and declining real price of oil
- 1973–1974 oil shock—Perceptions of rising demand and real price of oil, and some concern over the physical limit of oil resources
- 1975 recession—Perceptions of decline in demand and real price of oil, indefinite glut, and little concern over the future availability of oil
- 1979–1980 oil shock—Perceptions of rising demand for oil, large increase in the real price of oil, concern over political availability of oil and the physical resource base
- 1980–1983 recession—Perceptions of ever declining demand and real price of oil, little concern for the physical availability of oil, and expectation of indefinite oil glut
- 1984–1985—Slow revival of oil demand but no concern over availability, expectation of declining oil prices, and oil glut for many years
- 1986—Oil price crash and perception of total loss of OPEC power
- 1987–1988—Rebound of prices to $18 per barrel, no fear of crude availability in the market, and more confusion
- 1988 on—??

In each case, the immediate perception is projected up to 20 years ahead. This is why oil forecasting has become such a hazardous business. When confronted with their past mistaken projections, many forecasters put the blame on some unexpected OPEC or consumer policy change. Some groups, including the giant Exxon Corporation, no longer publish forecasts. The embarrassment of drastically wrong forecasts persuaded Exxon management to halt their publication.

Apart from the problem of perceptions, there are genuine difficulties in forecasting the demand and supply of oil. The major problems can be summarized as follows:

Demand

- What rates of gross domestic product (GDP) growth should be assumed, keeping in mind that each 0.5 percent difference in economic growth could lead to a 2 to 3 million barrels per day (b/d) difference in demand estimates for energy? That is, energy demand is extremely sensitive to GDP growth rates, and the rate of growth chosen for GDP significantly affects the rate of energy consumption.
- How will a change in capital stock affect the energy/GDP ratio? As the old, energy-inefficient machinery is replaced by new, energy-efficient machinery, the value of energy needed to generate one unit of GDP will decline. However, the turnover of machinery depends on interest rates and the general state of the economy. Ironically, in a recession many private firms cannot afford to pay for energy-efficient capital stock.
- How should the political will of governments to impose taxes or other administrative measures to reduce oil consumption be projected into the future? Different governments have different levels of political will to reduce dependence on oil. Because of election issues and domestic political crises, weak administrations are often reluctant to take measures that will raise the price of oil, even if the measures will ultimately curb oil consumption.
- What are the long-term price and income elasticities of demand for oil? While we can quickly measure the response of demand to price and income changes in the short run, we do not know what will happen in the long run. Perhaps by the mid–1990s we will have sufficient historical data to make such a judgment.

Supply

- How much oil is available both physically and technically? We need to ascertain the range of available oil supplies not only on the basis of total oil underground, but also in terms of the physical and technical availability. This requires detailed geological analysis and a good understanding of the political, legal, and engineering issues affecting availability.
- How much will the oil exporters prefer to produce? Oil exporters do not export as private firms, and oil resources are considered a national asset. Production policies are thus based on the long-term objectives of the country and the political impact of a particular course of action. Subjective, case-by-case analysis of the likely range of policies must be superimposed on the physical/technical availability of oil.

- What is the pace of development and quantity of exported non-OPEC oil? The growth of non-OPEC oil and its challenge to OPEC oil should be considered. How much investment is being made in non-OPEC oil, and can the growth of the 1980s be expected to continue? Will non-OPEC exporters begin to behave like OPEC after a certain period of time?
- What is the long-term price elasticity of oil supply? To what extent will the rise or fall in the price of oil affect the supply of oil from higher-cost areas?
- How will the investment in development of nonoil energy supplies be affected by general economic conditions (recessions and booms, inflation and interest rates, tax incentives) and by consumer governments' willingness to step in and directly help develop other energy sources?
- What will the cost of nonoil energy sources be in relation to oil? The cost of alternative energy supplies, particularly synthetic fuels, seems to rise with the price of oil. Synthetic fuels are projected to cost twice as much as oil. What is the realistic cost level of synthetic fuels that could be a ceiling for the price of oil?

There are simply too many imponderables in forecasting oil market developments. Oil forecasters can be blamed for shortsightedness and biases. Precise forecasts cannot and should not be expected.

PLANNING IN UNCERTAIN MARKETS

The key characteristic of today's oil market is volatility. Prices change suddenly, rising and falling in a pattern that is difficult to predict. When successful predictions can be made, they are usually made a week or a quarter ahead of time. In most oil companies, long-term planning has been put on the back burner in favor of taking advantage of short-term opportunities and avoiding short-term mistakes.

Some people may ask, "What's wrong with planning on a short-term basis?" For most companies, there is only the longer-term need for investment planning. Since there is a worldwide overcapacity of refining, petrochemicals, and tankers, there is little need for investment planning on the downstream side. For companies, the key issue is investment in exploration and production, and the recent downturn in these activities is symptomatic of uncertainty in the oil price outlook.

For governments, the problem is greater. Governments must make a wide variety of decisions that involve the longer-term picture. Regulations and policies must be adapted in light of long-term consequences; even the decision to abandon regulation altogether requires some sort of projection of market conditions over the long term.

The Changing Structure of the Oil Market

To understand the volatility of today's market, it may be useful to review the new forces that have come into play in recent years.

Prior to 1970, the oil market was fairly well controlled by the major international firms. They made most of the decisions affecting production, pricing, and destination of most oil exports. Posted prices ruled the day.

In the 1970s, OPEC took control of the market. Although OPEC generally responded only to the new realities in the market, following the increases in market price rather than setting it, OPEC official prices gave the market a solid and fairly reliable benchmark. During most of the 1970s and early 1980s, 90 to 95 percent of all oil exports were sold on contracts related to official prices.

It is instructive to compare today's situation with when the market was somewhat balanced, to see just how much has changed. Since the market was roughly in balance in 1971, and there were no major disruptions under way, the year 1971 will serve as the base year for comparison.

Excluding OPEC nations, every region in the world except North America produced more oil in 1985 than in 1971. Even North America's production was relatively steady. Between 1971 and 1985 non-OPEC oil production increased 59 percent. In 1971, OPEC produced about half of the world's oil. By 1985, OPEC was producing only 30 percent of the world's oil.

World oil consumption grew only 19 percent from 1971 to 1985. Dramatic changes took place where oil was demanded. Demand in OECD countries fell about 3 percent, while demand by non-OECD countries rose almost 71 percent. This means that OECD countries accounted for 70 percent of all oil consumption in 1971. The same countries account for only 57 percent of demand today.

With these factors in mind, the reasons for the recent oil glut become apparent: non-OPEC production rose by 59 percent during a period when oil demand rose by only 19 percent. This has led to a massive decrease in OPEC production. At the 1979 peak, OPEC was producing 31.5 million b/d; in 1985, this had fallen to 17.2 million b/d.

OPEC's new difficulties in selling oil have led to a major change in how oil is sold. Whereas about 90 to 95 percent of oil was sold on a contract basis at official prices just a few years ago, today only 10 to 15 percent is sold at official prices. Thirty to 35 percent is now moved directly on the spot market or is sold at prices corresponding to the spot market. The remaining 50 to 55 percent is sold at prices linked to the spot market, such as netback pricing, toll processing, and a host of other techniques that previously played a minor role. (See Table 1.1.) This new market structure has led to wild fluctuations in price. Although the feature that most people have noticed is the huge overall drop in the average price, the upward and downward fluctuations on a week-to-week basis are just as startling and unexpected.

Further confusion arises because of the futures market, a commodities market that trades for delivery at some future date. Although economic theory argues that futures markets tend to stabilize commodity markets by acting as a hedge against price fluctuations, the futures market in oil thus far has acted as a force that creates increased price volatility.

Table 1.1
Current Pricing Situation in the Crude Oil Market

30–35% of oil is sold directly on the spot market or at government sales prices corresponding to the spot market.

50–55% of oil is sold at prices linked to the spot market on a contract basis through mechanisms such as:

- netback pricing
- toll processing
- spot market escalators (formulas)
- production-sharing agreements
- countertrade (barter)

10–15% of oil is sold on contract at offical prices

In terms of the actual volumes of oil involved, the futures market is insignificant. Although approximately 45 million barrels of oil are traded on the New York Mercantile Exchange on a typical day, usually only about 700,000 to 800,000 actual barrels are sold (but they are sold repeatedly), thus skewing the statistics. The volumes of oil actually delivered from the futures market are small; in fact, in one case where a contract was due for delivery, it was found that there was not an actual cargo of oil available for delivery. Only two crudes are available for purchase on the futures market: West Texas intermediate (WTI) and North Sea brent. These represent a rather modest fraction of world crude supplies and are actually used in only a few areas of the world, especially since WTI cannot be exported from the United States. Total production of WTI is around 1.8 million b/d; the volume of WTI "traded" on the New York Exchange is thus 25 times the production volume, even though only about 10 percent of WTI production is actually sold on futures contracts. The large volume of transactions compared with actual oil production has led oil analysts to refer to such transactions as "paper oil."

Nevertheless, the futures market is important for two reasons. First, movements in the futures market tend to affect the current spot market, since they are supposedly an indicator of current oil industry thinking regarding the trend of prices in the near future. (In fact, most trading is done by speculators who know as much about oil as the average IBM stockholder knows about computers.) Second, it affects perceptions, because the news media find that the fluctuations in the futures market make for more dramatic news stories than the complex workings of the day-to-day oil market.

Although oil may fall to $10 per barrel on the futures market, this does not mean that the true price of oil has fallen. A common cry among refiners recently has been, "Well, where is this $10 oil I've been hearing about?" Even though the futures market price may drop to $10 per barrel, this is merely an option to take delivery at that price at some future date. Not only does this mean that the oil is not yet available at that price, but also that there is no guarantee that it will ever be acquired at that price. If the prices rise in the interim, the purchaser of the oil valued at $10 may resell it at $11; by the time it is actually delivered, the price could be dramatically different, and the same oil could have been resold literally thousands of times.

A question frequently raised is "How do you do market forecasting in today's market?" The answer, in short, is that it is not done—at least not in the traditional sense. Standard economic and modeling tools do not apply to today's situation. There is a wide gap between supply and demand. Domestic production is given priority by every country, which means that the marginal barrel of supply is usually cheaper to produce than the average barrel of supply. Under these conditions, there are no good equilibrium conditions to solve for.

Anyone who plans according to price forecasting in such an environment is either very brave or very foolish, or possibly both. The principal objective at present should be to determine the "outer bounds" (the ranges of likely outcomes) and how soon, or how much later, the market will return to a state where some of the standard variables will come back into play. Everyone knows that oil prices will eventually go up; the important question, then, is when prices will begin to rise again on a sustained basis.

What first must be determined is whether there are any scenarios that would result in a short-term sustained increase. It must also be determined whether non-OPEC producers will refrain from further cutbacks in support of the OPEC effort. China and the Soviet Union are unlikely to make major cuts because of their hard-currency needs. Egypt is having trouble marketing and would like to increase production. Mexico may cut around 150,000 b/d. Malaysia has promised a cutback of 10 percent, but this is a cutback from potential production and leaves actual production at its current level. Involuntary cutbacks of about 500,000 b/d have already taken place, mainly in the United States, Canada, and the North Sea. Thus, at best, no more than 500,000 b/d of voluntary cuts could take place, and even this is not assured.

A second consideration is the world political situation. Wars, revolutions, or other disruptions have to be evaluated. The largest exporter, Saudi Arabia, now produces only 4.3 million b/d. The second largest, Iran, is producing 2.3 million b/d. Shutting off all supplies from both of these countries for a prolonged period would certainly tighten the market for a while, but it would by no means create a supply emergency. It is difficult to create a reasonable scenario of supply cutoffs that would cause a sustained leap in the current market price; it is necessary to suppose some sort of prolonged disruption, such as war or natural catastrophe, involving a few major countries. No attempt to predict what would happen to prices under such conditions can be made here.

In the short term, then, it is impossible to find any highly probable factors arguing for a sustained price hike. Prices will probably firm, but it is difficult to identify anything that could push the price above $20 per barrel in real terms prior to 1990.

THE MEDIUM-TERM OUTLOOK

Three factors determine when prices firm and begin to rise: demand, non-OPEC production levels, and OPEC's required or desired levels of production.

Demand is of course the biggest factor. Each year, the Resource Systems Institute's Energy Program at the East-West Center in Honolulu conducts a survey of what everyone thinks about the future of demand—governments, industry, academia—and tries to get a feel for the realistic ranges. Of course, some of these forecasts are taken more seriously than others, depending on the methodology used, but there seem to be fairly reliable upper and lower limits. For the period through 1990, after compiling estimates for 11 regions of the world, it is hard to find realistic combinations that yield annual growth rates below 0.7 percent or above 1.2 percent, largely because of slow demand growth in OECD countries and Eastern Europe, which account for 77 percent of consumption. Even very rapid growth rates in the developing countries have little effect on raising this rate in the near term. For the 1990–1995 period, the rates that seem likely are 0.8 percent on the low side and up to 1.6 percent on the high side.

Turning to non-OPEC production, it is difficult to imagine a net increase between now and 1995. Although some increases will occur in some areas, they are more than likely to be offset by decreases in areas such as the United States and the North Sea. At best, non-OPEC production might be maintained at around 40 million b/d through 1995. Decreases seem more likely. Many companies that have done field-by-field analyses of their own reserves expect a loss of 1.5 million b/d by 1990, and a slightly greater loss between 1990 and 1995. Some of the more pessimistic forecasts predict a decline of 6 million b/d by 1995, twice as high as most expect. This would involve widespread shutdowns of existing production, including some fairly large fields. Such a scenario is rather unlikely, and a drop in non-OPEC production to about 37 million b/d by 1995 is much more probable.

Required or desired OPEC production levels present the greatest challenge, since each country must be reviewed in detail to determine production ceilings, optimal production levels from a geological point of view, income requirements, and political outlook. There are two questions to examine: What level of production would be regarded as representing some semblance of discipline? What level of production would the countries actually prefer? Looking first at estimates of preferred levels, in 1990 OPEC members would be satisfied producing about 23 million b/d; by 1995, 25.8 million b/d; and by 2000, 29.2 million b/d. (See Table 1.2.) Demand for OPEC oil above these levels would result in sustained hikes in the price of oil, putting it back on the path it seemed to be following in the 1970s.

For example, low demand and continuing high non-OPEC production would

Table 1.2
Preferred and Maximum Sustainable OPEC Production
(millions b/d)[a]

	1986			1990		2000	
	A	P	C	P	C	P	C
Iran	2.1	2.3	3.2	3.0	3.5	4.0	4.0
Iraq	1.7	2.0	2.0	3.0	4.0	4.0	6.0
Kuwait	1.3	1.3	2.2	1.3	2.2	2.0	3.0
Saudi Arabia	4.6	5.0	8.0	5.0	8.0	8.0	10.0
UAE	1.4	1.6	2.1	1.8	2.5	2.5	3.0
Qatar	0.3	0.5	0.5	0.5	0.5	0.3	0.3
Neutral Zone	0.3	0.5	0.6	0.5	0.5	0.3	0.3
Total Gulf	11.7	13.2	18.6	15.1	21.2	21.1	26.6
Indonesia	1.4	1.6	1.7	1.6	1.8	1.2	1.2
Libya	1.0	1.3	1.7	1.5	1.7	2.0	2.0
Ecuador	0.3	0.3	0.3	0.4	0.4	0.4	0.4
Venezuela	1.7	1.9	2.5	2.0	2.5	2.5	2.5
Gabon	0.2	0.2	0.2	0.2	0.2	0.2	0.2
Algeria	0.6	0.6	0.8	0.6	0.6	0.5	0.5
Nigeria	1.5	1.6	1.8	1.6	2.1	1.3	1.3
Total OPEC	18.4	20.7	27.6	23.0	30.5	29.2	34.7

a. Excluding natural gas liquids.

A = Actual January-July 1986.

P = Preferred/acceptable level of output.

C = Sustainable capacity (based on long-term production profiles).

Source: Resource Systems Institute, *Energy Program Files* (Honolulu: East-West Center, 1987).

represent the slackest market likely. In 1990 this scenario would indicate a demand for OPEC oil of around 20 million b/d, about 3 million b/d below OPEC's desired level. In 1995, it would result in 22.8 million b/d demand for OPEC oil—still leaving a difference of about 3 million b/d relative to what OPEC would like to produce.

High demand and large declines in non-OPEC production would represent the tightest market likely. In this case, demand for OPEC oil in 1990 would be about 22 million b/d—about 1 million b/d below OPEC's preferred level. Continued declines and continued high demand through 1995, however, could have an impressive effect; it would increase demand for OPEC production to 29 million b/d in 1995, with strong upward pressure on price.

About 21 million b/d of production would be necessary to maintain any effective OPEC discipline; discipline here means an end to violations of quotas, an end to widespread discounting by the larger producers, and a major curtailment of netback sales. Such a level of demand is not likely to occur before the end of the 1980s, according to the above figures; however, it seems likely that by the early 1990s, OPEC will again be in a position to act as a stabilizing force in pricing, even if major price increases cannot be forced.

Having looked at the range of possible outcomes, the likely scenario is herewith presented. This scenario is based on three assumptions:

1. The most important forces in the market react sluggishly and usually exert their influence when no one expects them.
2. History tends to repeat itself.
3. We don't learn from history.

In the likely scenario, demand grows at the lowest possible rate until 1990 and then surges to a high rate in the 1990–1995 period, after consumers have readjusted to lower prices. Non-OPEC production declines of 1.5 million b/d take place between now and 1990, but there are no declines after 1990 because of market firming. In any event, further declines are likely to be temporary shutdowns of wells rather than permanent closures.

This scenario gives an OPEC oil demand in 1990 of about 21.8 million b/d; by 1995, demand for OPEC oil will increase to 26.8 million b/d. This indicates a fairly rapid firming of prices between 1990 and 1995. It also gives history a chance to repeat itself. At 26.8 million b/d demand for OPEC oil, the situation recurs in which a revolution, war, or other disruption could easily result in another oil crisis. In 1995, OPEC production capacity is likely to be around 30 million b/d; there thus would be only 3.2 million b/d slack capacity between maximum physical capacity to produce and required production. In this situation, it is easy to imagine an instant replay of the oil shocks of 1973 and 1979 in 1995.

What does all this mean in terms of price? In real terms, it means that prices are unlikely to go above $20 per barrel before 1990. After 1990, however, market conditions could begin to tighten considerably, and much of OPEC's lost ground could be regained. It is reasonable to predict that between 1990 and 1995, real prices will be in the range of $20–26 per barrel. After 1995, real prices could rapidly resume their 1980 levels even with no disruptions, but political events after 1995 could easily result in exactly what has been experienced before: a sharp jump in prices and gradual erosion of established price levels as demand slowly responds. All the energy security and conservation considerations that many are now ignoring are likely to become key issues in the mid–1990s.

CONSUMER-PRODUCER DIALOGUE

A consumer-producer dialogue was taken seriously after the oil price increases in 1973, but has since undergone a series of changes. This has resulted in a greater separation between oil producers and consumers than was the case in the mid–1970s.

The First Experiment: The North-South Dialogue

The ill-fated Conference on International Economic Cooperation (CIEC), popularly known as the North-South Dialogue, was the first and last attempt at cooperation between oil producers and oil consumers. The need for such a dialogue arose from the first oil price shock of 1973–1974. There was much insecurity about supply availability and further oil price rises. Developing nations plunged into debt as they borrowed to pay for imports of high-priced oil. U.S. banks were busy in the Eurodollar market, recycling OPEC surplus revenues by lending to less developed countries with little regard for their creditworthiness.[2]

The industrial nations of the OECD were the initiators of the dialogue. They were seeking "secure" oil supplies and "reasonable" prices. For OPEC producers, this translated into making available all the oil OECD demanded at prices lower than they were willing to accept. In effect the oil exporters did not see any possible gain for themselves from this dialogue. Still, they agreed to negotiate, but with a provision: Other issues relating to developing countries should be discussed. OPEC insisted on the inclusion on the agenda of issues pertaining to finance, development aid, and technology transfer. After some disagreement, the industrial nations agreed to negotiate. The conference began in 1975 and dragged on for nearly two years.

It gradually became clear that the industrial nations were not prepared to give on anything, but instead were attempting to turn less developed countries against OPEC. The negotiations proved to be long, drawn out, and fruitless. In the view of this author, who had a role in the CIEC, the U.S. and West German governments wanted the conference to fail. At one point, a cable from Secretary of State Henry Kissinger to the U.S. ambassador at the conference was leaked to the press. The cable instructed the head of the U.S. delegation to keep OPEC busy negotiating during an OPEC ministers' conference, so that OPEC ministers would be too embarrassed to raise the prices while such vital negotiations were taking place.[3]

The failure of the CIEC was a major blow to the chances for successful, long-term oil producer-consumer dialogue. Oil producers felt that the industrial countries were not willing to give up anything. They in turn did not concede much. There was a good possibility that an agreement could have been reached if the industrial nations were really interested. OPEC members at the time were so nervous about their own actions that they might have made major concessions. The failure of the CIEC created a feeling of mutual distrust. It set back the opportunity for future dialogue for a number of years.

Factors Affecting the Dialogue

After the CIEC and the tightening of the oil market that resulted from the Iranian Revolution, many OPEC countries believed they had complete control over pricing. They felt negotiation with consumer governments was unnecessary because they operated on the basis of the same perceptions as the Western world: Oil prices were expected to rise unchecked. At the time, consumer governments were eager to engage OPEC in dialogue. With the decline in oil demand and the weakening of prices, the attitudes somewhat reversed. OPEC countries that felt the pain of lower oil prices were eager to solicit the assistance of consumer governments in stabilizing prices. Both sides shifted positions, depending on the circumstances, but neither side displayed the foresight needed for real dialogue.

Three consumer countries deserve special mention with regard to the developing discussions.

- *United States.* Since the beginning of the Reagan presidency, the administration has taken the position that any dialogue or discussion that "might" lead to an impact on oil prices (which would presumably be the reason for the discussion) would be unacceptable—since it would interfere with the free market. On the other hand, discussions about the security of supply would be welcome. Clearly, the issues of price and security of supply are inseparable, and no oil producer will discuss one without the other. The administration has banned participation of its staff even in conferences where the issues might arise. Given the U.S. influence on many major consumers and the International Energy Agency, this position has significantly hampered any dialogue. Notwithstanding the U.S. government's position, Vice-President Bush did go to Saudi Arabia in 1986 to "cordially request" the Saudis to raise the price of oil to help his fellow Texans.
- *Japan.* Japan has always expressed sentiments that a dialogue between producers and consumers should take place, although a concrete plan has never been set forth. As a resource-poor OECD member, Japan is acutely aware of the importance of being at least in the good graces of OPEC countries. Having established a good working relationship, there is little wonder that Japan is not interested in employing large naval forces to protect the oil lanes of the Persian Gulf. Its most important contribution to the producer-consumer relationship came in the wake of the 1986 collapse of oil prices. Japan seriously feared that lower oil prices might endanger the political stability of the oil exporters. Emphasizing the philosophy that oil importers should not take advantage of the temporary weakness in prices, the Japanese proposed a compromise oil price of $18 per barrel. Though other industrial governments did not publicly endorse the idea, this proposition won support among many oil producers and consumers. When OPEC chose to push prices up to $18 per barrel in late 1986–early 1987, Japan's blessing provided important moral and political support.
- *Italy.* Successive Italian governments have encouraged close relations and dialogue between consumers and producers of oil. Unlike the Japanese, the Italians did not volunteer to pay high prices but instead emphasized "forward integration" of the oil producers into consumer countries. This meant encouraging oil producers to build refining and petrochemical industries in oil-consuming countries. The Italians felt this activity, financed by oil exporters, would commit both the oil and the capital of OPEC

countries in the oil-consuming areas. The state oil company, ENI, undertook a series of in-depth studies with the Organization of Arab Petroleum Exporting Countries to show the value of this type of investment.

Having discussed the role of these key oil-consuming governments, it is worthwhile to mention the oil companies' interest in such dialogue. At the outset, it is important to emphasize what matters most to the oil companies: profit. Profit in the oil business is the difference between acquisition cost (producing crude or buying crude) and the sale price of refined products. All that the companies care about is their profit margin. While stable prices might lead to positive margins, only uncertainty and wide fluctuations can lead to substantial profits. For the companies engaged primarily in exploration and production, high and stable prices are extremely important. These differing goals often lead to conflicts among production and marketing subsidiaries of the same company. Fundamentally, however, oil companies do not now have an important desire or a major say in whether a consumer-producer dialogue should take place.

THE FUTURE

The unmistakable fact about oil that many people tend to forget is that it is an exhaustible resource. U.S. oil production will decline every year no matter what government policies are in effect or what the price is. The rate of decline may decrease, but the trend is unchangeable. Like it or not, the geologic fact is that most oil resources are concentrated in Iran, Iraq, Saudi Arabia, Kuwait, the United Arab Emirates, Venezuela, and Libya. As the end of the century approaches, these producers will have even greater control over oil supplies than OPEC did in the 1970s. Will they learn from history and keep prices at moderate levels? Probably not. Oil is an important component of the political and social fabric of these countries, and any indication that prices are deliberately kept below what the market will bear will be politically suicidal for the parties involved.

It seems that the opportunity for effective producer-consumer dialogue has been reduced to almost nil. Not only have both sides failed to come up with a workable plan in the past, but it may now be too late for any kind of plan in the future. There are too many players in the oil market: from the "Wall Street refiner" to speculators on the commodity exchanges to virtually thousands of middlemen who trade oil. Even if the oil producers can keep prices below the market clearing prices, the speculators and middlemen will raise the prices and reap the benefits.

The only serious prospect for consumer-producer dialogue is further emphasis on and expansion of the Italian plan for forward integration. Consumer governments should consider actively encouraging OPEC nations to invest in the oil industry. It is much more difficult to shut down refineries in a consuming market with millions of consumers than to shut off the crude valve. Not only will OPEC nations risk seizure of assets in the event of confrontation, but they will genuinely learn about the complexities of survival at the retail level, thereby significantly reducing the prospects of a political cutoff of oil supplies. So far only Kuwait (in Europe) and Venezuela (in the United States and Europe) have ventured into downstream markets. Other countries are hesitant, fearing political backlash from consumer governments. This is where there is some scope for improvement. Of course, forward integration is only a minor step on the road to consumer-producer dialogue. But it is better than nothing!

NOTES

1. See F. Fesharaki and S. Hoffman, "Medium-Term and Long-Term Outlook for Oil: Survey and Analysis of the State of the Art," Resource Systems Institute, East-West Center, Honolulu, 1982; *Energy: The International Journal* (1983).

2. The debt crises of the developing nations today are the direct result of the excessive lending policies by U.S. banks in the mid–1970s.

3. R. Fallah and F. Fesharaki, "The New World Economic Order: The Energy Dimension," *Middle East Economic Survey* 20, no. 3 (June 1977).

TWO

The Energy Sector and Economic Growth

WEN S. CHERN

INTRODUCTION

Many developing countries, particularly those in the Pacific Rim area, have maintained relatively high rates of growth in both their economy and their consumption of commercial energy sources (including petroleum, coal, natural gas, and hydro power) since the 1973 oil embargo. The relatively high rates of growth in energy demand have caused energy analysts to be concerned about the likely increasing role of developing countries in the future world energy market. Clearly, the developing countries need energy resources for economic development. But can these countries produce additional units of goods and services with smaller amounts of energy, as many industrialized nations have done? Unfortunately, there is no simple answer to this question, because most developing countries need further industrialization in order to sustain their economic development. Also, a large fraction of their energy requirements is met by conventional noncommercial energy sources, such as firewood, charcoal, and biomass. As economic development proceeds, these noncommercial energy sources are likely to be replaced by commercial energy sources, thus pushing up the demand for the latter.

The purpose of this chapter is to analyze the linkages between the energy sector and economic growth, and the factors affecting energy demand in South Korea, the Philippines, Singapore, Taiwan, and Thailand. Specifically, it provides a descriptive analysis of the relationship between energy and economic development, and a statistical estimation of income and price elasticities of energy and oil demand in these five Asian oil-importing countries. It is motivated by a growing concern about the changing structure of energy and oil demand in less-developed countries (LDCs). The emerging uncertainties for assessing the energy-economy relationships are attributed to the following:

- The dramatic declines in oil prices in 1986, resulting from OPEC's failure to control its production, the dramatic expansion of non-OPEC production capacity, and the subsequent volatility of world oil prices
- The slowdown of growth in both aggregate energy and oil consumption in 1980–1984.

In particular, the growth in both total and per capita oil consumption has either significantly slowed or become negative during the 1980s in these five Asian countries. The data show that energy and oil consumption in most LDCs continued to grow at fairly rapid rates after the oil shocks of 1973 and 1979. In view of these recent events, the critical question is whether the demand for energy and oil will slow in the LDCs. In order to answer this question, statistical evidence on the estimated income and price elasticities of energy and oil demand will be developed and used.

ENERGY AS A SOURCE OF ECONOMIC DEVELOPMENT

The fact that energy is an important factor of production has been widely investigated and documented since the oil embargo. At the aggregate level, a nation's total production of goods and services may be measured by gross domestic product (GDP) or gross national product (GNP). A strong relationship between energy and GNP was shown in earlier studies by Hudson and Jorgensen (1974) and the Institute for Energy Analysis (1979). At the sectoral level, the output of a sector may be measured by value added, gross output, or a production index. The impacts of energy on sectoral output have been statistically shown in numerous studies, including those by Berndt and Wood (1975, 1979) and Griffin and Gregory (1976). One focus of these studies was the substitutability between energy and other factors of production. Energy and labor were found to be substitutes. However, conflicting results were found regarding relationships between energy and capital. If energy and capital are complements, as shown by Berndt and Wood, then subsidizing capital equipment may be counterproductive for energy conservation. Of course, other reasons, such as speeding up industrialization, may override this consideration.

The importance of energy as a factor of production does not imply constancy of the energy-output relationship. Although energy is required to produce any good, the amount of energy required for one unit of output may be changed. When energy was plentiful and relatively inexpensive during the 1950s and 1960s, additional energy was used to substitute for other factors of production in most industrial and household production. Briggs and Borg (1986) showed that the energy/GNP ratio has steadily declined since 1970 in the United States. This dramatic trend reflects the effects of the energy conservation efforts made by consumers in response to the two oil shocks in the 1970s.

The basic relationship between energy and economic growth applies to LDCs even though the specific relationship depends upon their degree of economic development. Table 2.1 shows the correlation coefficients between per capita

Table 2.1
Correlation Coefficients Between Per Capita Real GDP and Per Capita Energy Consumption

	Per capita real GDP and per capita energy consumption		Per capita real GDP and per capita oil consumption	
Country	1960–73	1974–84	1960–73	1974–84
Philippines	0.97	0.53	0.97	0.17
Singapore	0.97	0.95	0.97	0.95
South Korea	0.97	0.95	0.99	0.90
Taiwan	0.99	0.94	0.99	0.70
Thailand	0.97	0.96	0.97	0.67

Sources: Council for Economic Planning and Development, Republic of China, *Taiwan Statistical Data Book* (Taipei: The Council, various issues); Energy Committee, Republic of China, *Taiwan Energy Statistics* (Taipei: The Committee, 1980, 1984, 1985 issues); United Nations, *Yearbook of World Energy Statistics* (New York: United Nations, 1980, 1984, 1985 issues); World Bank, *World Tables* (Washington, D.C.: World Bank, 1976, 1980, 1983, 1986 issues).

real GDP and per capita energy consumption (and a corresponding set for per capita oil consumption) in the five Asian countries. The correlation coefficients were computed for two periods for the purpose of comparison before and after the 1973 oil embargo.

Consider first the correlations between per capita real GDP and per capita energy consumption. High correlations are observed in all countries. In the Philippines, the correlation coefficient in 1974–1984 is considerably smaller than that in 1960–1973. But, by and large, a strong correlation persists in the four other countries, implying that energy remained an important factor for GDP growth in these developing countries after the 1973 oil embargo. The correlation coefficients between GDP and oil consumption were generally very high during 1960–1973. However, in the Philippines, Taiwan, and Thailand, the correlations became smaller after the embargo, suggesting possible structural changes in the oil/GDP relationships in these countries. In fact, the correlation coefficient for the Philippines is unreasonably small for 1974–1984. Apparently, this small correlation is largely due to the volatility of both oil consumption and per capita GDP during this period. The correlations remain very high in South Korea and Singapore. The latter is noted for its almost complete dependence on imported oil and its huge refining sector.

Historical trends in energy/GDP and oil/GDP ratios in Asian LDCs are shown in Table 2.2, which presents indices of the average ratios for five partitioned periods during 1960–1984. The base period is 1960–1984. The energy/GDP ratio reflects the intensity of energy use for the aggregate economy. While this ratio

Table 2.2
Indices of Average Energy/GDP and Oil/GDP Ratios by Country (base period, 1960–64)

Item/Country	1960–64	1965–69	1970–73	1974–79	1980–84
Index of average energy/GDP ratio					
Philippines	100	144	164	149	129
Singapore	100	106	96	72	56
South Korea	100	132	159	152	170
Taiwan	100	100	102	114	116
Thailand	100	154	204	214	202
Index of average oil/GDP ratio					
Philippines	100	150	174	157	118
Singapore	100	106	96	72	56
South Korea	100	259	501	506	514
Taiwan	100	175	291	378	335
Thailand	100	154	191	198	171

Sources: Council for Economic Planning and Development, Republic of China, *Taiwan Statistical Data Book* (Taipei: The Council, various issues); Energy Committee, Republic of China, *Taiwan Energy Statistics* (Taipei: The Committee, 1980, 1984, 1985 issues); United Nations, *Yearbook of World Energy Statistics* (New York: United Nations, 1980, 1984, 1985 issues); World Bank, *World Tables* (Washington, D.C.: World Bank, 1976, 1980, 1983, 1986 issues).

may be appropriately employed to measure the trend of aggregate energy utilization efficiency in industrialized nations, it may reflect more of the changes in economic structure in most LDCs (agriculture vs. industry, or industry vs. service sector). The historical trends show that the index in 1980–1984 is higher than in 1960–1964 in all countries except Singapore. In fact, the computed indices show strong evidence that all these countries, except Singapore, moved to a more energy-intensive economy after the 1973 oil embargo. The most dramatic increases in the energy/GDP ratio are observed in Korea and Thailand. In Singapore, the trend is similar to those observed in the United States and other industrialized nations.

Table 2.2 also shows indices of the oil/GDP ratio for the selected periods. The role of oil in economic development is reflected by dramatic increases in the oil/GDP ratio in Korea, Thailand, and Taiwan. The index decreased in 1980–1984 from the preceding period in all countries except Korea, apparently a result of the efforts to reduce foreign oil dependence after the two oil shocks. However, in Korea the average indices of energy/GDP and oil/GDP ratios were higher in 1980–1984 than 1974–1979. These ratios peaked in 1980, and since then the trends have been declining.

DIVERSIFICATION OF ENERGY SUPPLY SOURCES

Since the oil embargo in 1973, most developing countries have followed policies like those of the industrialized nations to diversify the sources of energy supply. The strategies for diversification take many forms, depending upon the endowment of various energy resources, economic conditions, and other environmental political considerations. For example, in Korea and Taiwan the strategy has been to develop nuclear power and to increase use of coal and LNG (liquefied natural gas). In the Philippines, development of hydro and geothermal power has been the main approach to diversification of energy supplies. In Thailand, the national strategy is based on the potential of domestic natural gas and coal, and imported coal.

In general, reduction of foreign oil dependence has been a national goal for most Asian developing countries. The need for reducing such foreign dependence is obvious. Since 1973 the energy import bill has skyrocketed as a result of the massive oil price increase during the two oil shocks. The balance-of-payments effects of raising oil prices were serious in all five Asian countries under investigation. Specifically, in 1974 the cost of imported oil rose (in current prices) from 1973 by U.S.$700 million for South Korea, U.S.$400 million for Singapore, U.S.$500 million for the Philippines, U.S.$400 million for Thailand, and U.S.$616 million for Taiwan. On the average, the increment in the oil import bill exceeds 2 percent of GDP in all these countries. The impact of the second oil shock was even more severe. In 1979 and 1980 South Korea's oil import bill rose by a cumulative figure of nearly U.S.$4 billion, Singapore's by U.S.$600 million, the Philippines' by more than U.S.$800 million, Thailand's by U.S.$1.7 billion, and Taiwan's by U.S.$3.1 billion.

Table 2.3 shows the net energy imports and their shares of total energy consumption for 1971 and 1984. Net energy imports increased substantially in all five Asian countries. The net import shares of total energy consumption measure the degree of dependence on foreign sources of energy (primarily oil, but also coal, LNG, and nuclear power). The figures in Table 2.3 indicate that the successes in reducing foreign dependence rely on the availability of domestic energy sources. In Korea and Taiwan, import shares have continued to rise from 1971 until recent years. Both countries imported less oil in 1984, and the trend continued at least through mid–1986, partially due to national policies aimed at reducing energy waste. Both countries treat nuclear power as imported energy. Singapore has had a virtually total dependence on foreign energy. In the Philippines and Thailand, import shares have declined, suggesting a reduced foreign energy dependence in recent years.

Diversification of energy supply should result in changes in fuel mix. Table 2.4 shows total primary energy consumption and fuel shares for the selected five Asian LDCs. Total primary energy consumption continued to increase very rapidly from 1971 to 1984. In terms of the fuel mix, the data clearly suggest that despite diversification efforts, oil continued to be the dominant fuel in all

Table 2.3
Net Energy Import Quantities and Shares

Country	Year	Net energy imports (million toe)	Net import shares (%) of total energy consumption
Philippines	1971	8.5	87.6
	1984	10.2	70.8
Singapore	1971	6.6	100.0
	1984	11.3	100.0
South Korea	1971	11.1	63.4
	1984	34.8	70.3
Taiwan	1971	6.6	70.2
	1984	27.0	95.1
Thailand	1971	6.1	92.4
	1984	10.8	67.5

toe = Metric tons of oil equivalent.

Sources: Council for Economic Planning and Development, Republic of China, *Taiwan Statistical Data Book* (Taipei: The Council, various issues); Energy Committee, Republic of China, *Taiwan Energy Statistics* (Taipei: The Committee, 1980, 1984, 1985 issues); United Nations, *Yearbook of World Energy Statistics* (New York: United Nations, 1980, 1984, 1985 issues); World Bank, *World Tables* (Washington, D.C.: World Bank, 1976, 1980, 1983, 1986 issues).

of the countries. However, the market shares of oil declined in all countries except Singapore, where its primary energy source remained 100 percent oil. Natural gas accounted for a very small market share in most countries except Thailand, where the increase from 1971 to 1984 was substantial. Coal accounted for a significant market share in Korea and Taiwan. In Taiwan, coal has been mainly imported in recent years.

The shares of hydro and nuclear power have increased substantially in Korea, the Philippines, and Taiwan. In Taiwan and Korea, nuclear power accounted for most of the increases. In the Philippines, the development of geothermal and hydro power has been more successful than the expansion of natural gas and coal use for diversification.

The preceding analysis suggests that four of the five Asian countries (Korea, the Philippines, Taiwan, and Thailand) have achieved considerable success in reducing their foreign oil dependence. However, trend analysis indicates that these countries will continue to depend on oil for their economic development for many years to come.

ESTIMATION OF INCOME AND PRICE ELASTICITIES OF ENERGY AND OIL DEMAND

The model used in this study represents an extension of the model used by Chern (1987) and Chern and Soberon-Ferrer (1987). Basically, this is a dynamic

Table 2.4
Total Primary Energy Consumption and Fuel Shares

Country	Year	Total primary energy consumption (million toe)	Shares (%) of Oil	Natural gas	Coal	Hydro and nuclear power
Philippines	1971	9.7	94.8	—	—	5.2
	1984	14.4	73.0	—	4.8	22.2
Singapore	1971	6.8	100.0	—	—	—
	1984	10.9	100.0	—	—	—
South Korea	1971	17.5	62.9	—	34.9	2.2
	1984	49.5	56.8	—	36.0	7.2
Taiwan	1971	9.4	60.6	7.4	25.5	6.5
	1984	28.4	53.9	3.9	21.8	20.4
Thailand	1971	6.6	90.8	—	1.5	7.7
	1984	16.0	74.4	13.1	5.0	7.5

toe = Metric tons of oil equivalent.

— = Insignificant shares.

Sources: Council for Economic Planning and Development, Republic of China, *Taiwan Statistical Data Book* (Taipei: The Council, various issues); Energy Committee, Republic of China, *Taiwan Energy Statistics* (Taipei: The Committee, 1980, 1984, 1985 issues); United Nations, *Yearbook of World Energy Statistics* (New York: United Nations, 1980, 1984, 1985 issues); World Bank, *World Tables* (Washington, D.C.: World Bank, 1976, 1980, 1983, 1986 issues).

demand model with the partial adjustment specification and the pooling of time-series and cross-section data. For estimating aggregate energy demand, the model is specified as follows:

$$\ln ENE_{it} = f(\ln ENE_{i,t-1}, \ln REP_{it}, \ln RGDP_{it}, \ln SMAF_{it}, SEIM_{it}, D_1 \ldots D_4, \epsilon_{it}) \quad (1)$$

Where:

i = country

t = year

ENE = per capita consumption of energy (toe/person)

REP = weighted average of energy prices (U.S.$/toe)

RGDP = real per capita GDP (1975 U.S.$)

SMAF = share of GDP in the manufacturing sector

SEIM = share of energy imports in total energy consumption

D = country dummy variables (D_1 = Korea, D_2 = Philippines, D_3 = Singapore, D_4 = Thailand, Base Country = Taiwan)

ϵ = disturbance term.

The inclusion of country dummy variables depends upon the estimation method (to be discussed later).

For estimating oil demand, the dependent and energy price variables are replaced, respectively, by

OIL = per capita consumption of oil (toe/person)
ROP = weighted average price of oil products (U.S.$/toe).

It is recognized that there is a methodological problem of modeling aggregate energy and oil demands independently. If we use the estimated equations for forecasting, there is no assurance (that is, constraint) that oil demand is always smaller than the aggregate demand. It would have been more appropriate to adopt a logit model specification in order to estimate aggregate energy demand and the share of oil. Unfortunately, data limitations have prevented us from taking this approach. For Thailand and the Philippines, we do not have sufficient data on the prices of natural gas, coal, and electricity to construct the weighted average price variable, REP, in Equation (1). For these countries, the weighted average price of oil products, ROP, is used as an approximation. In order to estimate a logit model for oil share, we need complete price data on oil versus all other energy sources.

Two methods are considered for estimating the aggregate energy and oil demand equations. The first method assumes the following error structure:

$$\text{COV}\ (\epsilon_{it}, \epsilon_{jt}) = \begin{Bmatrix} \sigma_1^2 \text{ if } i = j \\ \text{O otherwise} \end{Bmatrix}$$

Therefore, the variance differs from country to country (heteroscedasticity). Under this specification, we also include the country dummy variables in the structural equations. The appropriate method is then to extend ordinary least squares with dummy variables to weighted least squares (WLS). There is one more complication. The variable SEIM is a ratio of energy imports to total energy consumption. Therefore, it is not entirely exogenous (total energy consumption is part of the dependent variable). Hence, we need to incorporate the instrumental variable procedure to obtain an estimate of SEIM as an exogenous variable before implementing WLS. We term this first method instrumental variable weighted least squares (IWLS).

The second estimation method adopts a specification of the error component model developed by Wallace and Hussain (1969), Maddala (1971), and Fuller and Battese (1974). Specifically, the error structure is specified as follows:

$$\epsilon_{it} = \mu_i + v_t + e_{it}$$

where μ_i, v_t, and e_{it} are independently distributed with zero means and variances σ_μ^2, σ_v^2, and σ_e^2, respectively. Therefore, the error term is decomposed into three components. μ_i is an unobserved effect for the i-th country, v_t is an unobserved effect for the t-th year, and e_{it} is an observed remainder that differs among countries as well as time periods. The computation method is based on the procedure developed by Fuller and Battese. The complication involving SEIM applies here also. Consequently, the instrumental variable procedure is first applied to obtain an estimate of SEIM before implementing Fuller and Battese's generalized least squares (GLS). We term this second method instrumental variable generalized least squares (IGLS).

Data on primary energy and oil consumption, and energy imports and exports, were obtained from the World Bank. The World Bank's data were provided by the United Nations with a special computer tape. All energy units in this data base were in metric tons of oil equivalent (toe) instead of the metric tons of coal equivalent used in United Nations' *World Energy Supplies* and *Yearbook of World Energy Statistics*. The World Bank data base does not include Taiwan. For Taiwan, data were obtained from *Taiwan Energy Statistics*, published by the Energy Committee. All macroeconomic and demographic data, including GDP (total and sectoral), GDP deflator, population, and exports and imports, were based on the *World Tape*, released by the World Bank for all countries except Taiwan. The macroeconomic and demographic data for Taiwan were obtained from the *Taiwan Statistical Data Book*, published by the Council for Economic Planning and Development. Other statistical sources of data include the United Nations' *Statistical Yearbook* and *Demographic Yearbook*. Data on energy prices were obtained from diverse sources, mostly from these countries directly. The sample period covers 1970–1984.

The regression results are presented in Table 2.5. They show that per capita GDP and energy price (oil price) are the most significant determinants of energy demand (oil demand). The estimated coefficients of the lagged dependent variable are all plausible (being between the theoretical range of zero and one). The estimated coefficients of *ln* SMAF, the GDP share in the manufacturing sector, all have a positive sign, but the estimated standard errors are relatively high (*t*-ratios are less than 2.0). The foreign energy dependence variable, SEIM, is not statistically significant. This result may be due to the lack of sufficient variation because all five countries had a very high share of imported energy during the sample period. This variable was found to be significant in the model estimated for a more diverse group of LDCs by Chern (1987). The estimated coefficients of the country dummy variables under IWLS show that the unaccounted differences in aggregate energy demand are not significant among these five countries. However, for oil demand, the estimated country dummy coefficients show that the Philippines tended to use more oil, and Singapore less, than the other three countries while holding other factors constant.

Both estimation methods produce similar and plausible results in terms of the sign and *t*-ratios. Based on these parameter estimates, it would be difficult to make a judgment on the superiority of either. IGLS has a much higher SSE than IWLS, but this is typical because the country dummy variables under IWLS always enhance the explanatory power.

Table 2.5
Regression Results[a]

	Aggregate energy demand (ln ENE)		Oil demand (ln oil)	
Exogenous variables	IWLS	IGLS	IWLS	IGLS
Intercept	0.942	0.857	0.860	0.968
	(2.98)	(3.50)	(2.54)	(5.02)
ENE_{t-1}	0.825	0.870		
	(13.66)	(20.18)		
ln OIL_{t-1}			0.702	0.789
			(12.50)	(17.38)
ln REP[b]	-0.116	-0.106	-0.139	-0.163
	(-3.72)	(-4.07)	(-4.11)	(-6.07)
ln RGDP	-0.169	0.143	0.173	0.281
	(2.54)	(2.36)	(2.39)	(4.38)
ln SMAF	0.196	0.162	0.286	0.074
	(1.39)	(1.63)	(1.90)	(0.97)
SEIM	-0.083	-0.095	0.151	-0.100
	(-0.55)	(-0.95)	(0.85)	(-1.01)
D_1 (Korea)	0.012		-0.007	
	(0.37)		(-0.21)	
D_2 (Philippines)	-0.102		-0.192	
	(-1.18)		(2.52)	
D_3 (Singapore)	0.041		0.290	
	(0.40)		(2.52)	
D_4 (Thailand)	-0.015		-0.100	
	(-0.14)		(-1.022)	
SSE	0.0105	0.159	0.0129	0.209
R^2	0.9978		0.998	

a. Figures in parentheses are asymptote *t*-ratios; SSE is the sum of squared errors; R^2 is the coefficient of determination and is not computed under IGLS.

b. For Thailand and the Philippines in equation 1, the weighted average price of oil products was used as a proxy.

Sources: Council for Economic Planning and Development, Republic of China, *Taiwan Statistical Data Book* (Taipei: The Council, various issues); Energy Committee, Republic of China, *Taiwan Energy Statistics* (Taipei: The Committee, 1980, 1984, 1985 issues); United Nations, *Yearbook of World Energy Statistics* (New York: United Nations, 1980, 1984, 1985 issues); World Bank, *World Tables* (Washington, D.C.: World Bank, 1976, 1980, 1983, 1986 issues).

Table 2.6
Estimated Income and Price Elasticities for Five Asian Countries

Model and estimation method	Income elasticities		Price elasticities	
	Short-run	Long-run	Short-run	Long-run
Aggregate energy demand				
IWLS	0.169	0.966	-0.116	-0.663
IGLS	0.143	1.100	-0.106	-0.815
Oil demand				
IWLS	0.173	0.581	-0.139	-0.466
IGLS	0.281	1.332	-0.163	-0.773

Sources: Council for Economic Planning and Development, Republic of China, *Taiwan Statistical Data Book* (Taipei: The Council, various issues); Energy Committee, Republic of China, *Taiwan Energy Statistics* (Taipei: The Committee, 1980, 1984, 1985 issues); United Nations, *Yearbook of World Energy Statistics* (New York: United Nations, 1980, 1984, 1985 issues); World Bank, *World Tables* (Washington, D.C.: World Bank, 1976, 1980, 1983, 1986 issues).

Table 2.6 shows the estimated short- and long-run income and price elasticities of energy and oil demand for these five Asian countries. In comparing the estimated elasticities, there are significant differences between the two estimation methods, with IWLS producing smaller income and price (in absolute value) elasticities than IGLS. In particular, the estimated long-run income elasticity of oil demand is 0.581 under IWLS and 1.33 under IGLS. This dramatic difference suggests the instability of oil demand in recent years. The estimate of 1.33 under IGLS is unreasonably large. This estimate may be biased upward because the specification of the error components model does not allow different constant terms for different countries. Specifically, if there are unexplained differences in oil use patterns among countries (and these may result from different government policies), as may have occurred after the second oil shock of 1979, the estimated income elasticity may be disturbed simply because of no allowance for these differences in the model. The estimated income elasticities were much more robust in our previous studies with data prior to 1982. The statistical results provide strong evidence about the likelihood of declining income elasticities of aggregate energy demand, particularly oil demand in recent years. However, the structural changes are by no means stable. It would be essential to update this model with additional data (say 1985–1986) in order to trace these changes more definitely.

The estimated price elasticities in absolute value are all higher than those found in previous studies for LDCs. For example, the long-run price elasticities estimated by IGLS are -0.815 for aggregate demand and -0.773 for oil demand. These results strongly suggest that the price responsiveness of energy and oil demand has become much stronger in the five countries in recent years. The higher price elasticities, however, would add another dimension of uncertainty in forecasting energy and oil demand under unstable energy and oil prices.

CONCLUSIONS

This study provides descriptive statistics of the relationship between energy and economic development, and econometric estimates of income and price elasticities of aggregate energy and oil demand for five Asian developing countries. The important findings of this statistical analysis are summarized below.

First, a strong relationship between energy use and GDP growth continued after the oil shocks in 1973 and 1979. The oil/GDP ratio declined during 1980–1984 in all countries except Korea. (In Korea, the oil/GDP ratio peaked in 1980 and has been declining ever since.) These declines in the oil/GDP ratio apparently resulted from the efforts to reduce foreign oil dependence after the two oil shocks.

Second, despite the efforts by all these countries (except Singapore) to diversify the sources of energy supply and despite the declines in oil/GDP ratios during 1980–1984, oil continued to be the dominant fuel in 1984. The available data still suggest that these oil-importing countries remain highly vulnerable under any oil supply disruption.

Third, the statistical estimation shows that the aggregate oil demand may have undergone structural changes seen in the falling income elasticity in recent years. However, the results show that the extent of structural changes cannot be determined without additional data because the income elasticities are highly sensitive to alternative estimation methods.

THREE

Energy Trade in Asia and the Pacific

TORU KIMURA

POSITIONING OF THE PACIFIC REGION

Figure 3.1 shows the position that the Pacific region held in world energy supply and demand during 1985. It has a mere 24.4 percent of total oil reserves in the world. The region produces 30.6 percent of world oil output and uses as much as 41.1 percent of world oil consumption, thus revealing its heavy weight in consumption despite its poor resource availability. As a result, it takes 34.3 percent of total oil imports in the world. This forms a striking contrast with its share of oil exports, standing at 17.4 percent, and suggests large amounts of oil flowing into the Pacific region from outside. As for natural gas, the shares of production (26.9 percent) and consumption (23.2 percent) of North America are much higher than its share of reserves (16.4 percent). Small differences are seen in the shares (reserves, 5.6 percent; production, 5.0 percent; and consumption, 5.8 percent) of Asia and Oceania,[1] where the development and use of natural gas imports and exports are 16.7 and 14.7 percent, respectively, suggesting that natural gas trade unfolds largely within the Pacific region. In the case of coal, the shares of production (26.0 percent) and consumption (25.8 percent) of the planned economies (mainly China) outpace their share of reserves (19.2 percent), but are well balanced as a whole. Compared with oil reserves (24.4 percent) and natural gas reserves (20.3 percent), coal reserves (51.3 percent) are much more abundant in the Pacific region. Unlike oil, the share of coal exports (57.6 percent) in the Pacific region is higher than its share of coal imports (40.3 percent). In addition, the region contains as much as 43 percent of available hydro resources in the world.

Thus, one of the characteristics of energy supply and demand in the Pacific region is the scarcity of oil resources. Excessive oil dependence in the region

Figure 3.1
Position of the Pacific Region in World Energy Supply-Demand

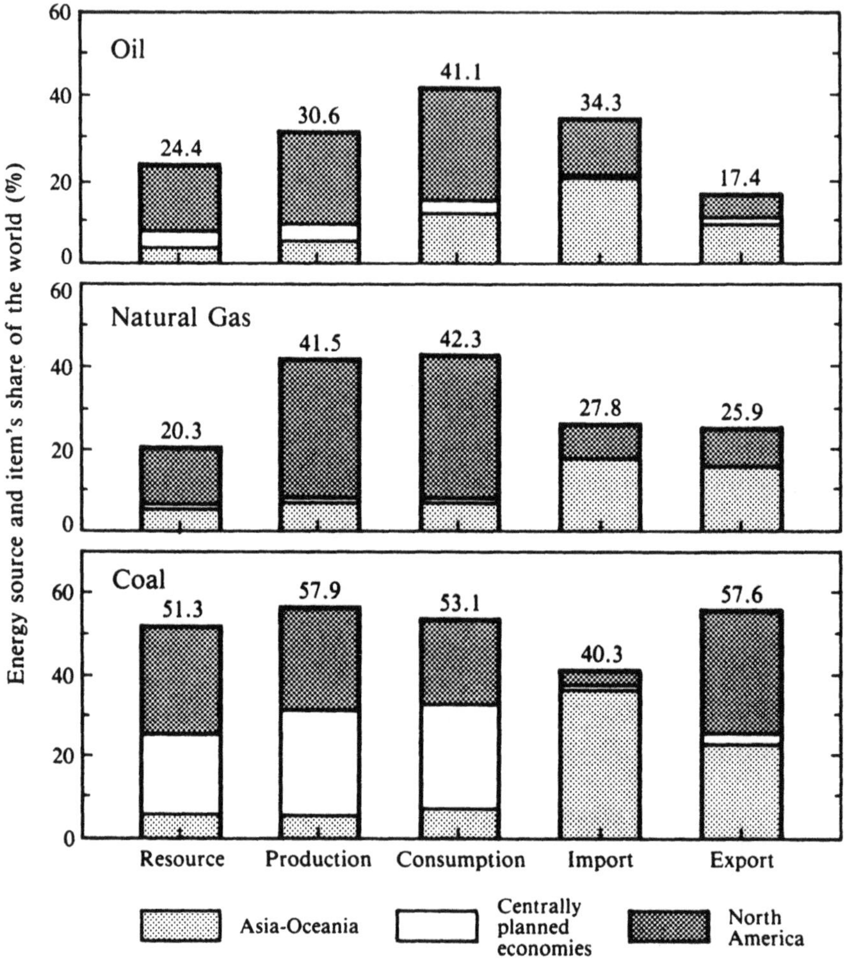

Source: United Nations, *Yearbook of World Energy Statistics*, (New York: United Nations, 1985).

not only makes the energy supply and demand structure inflexible, but also renders supply security vulnerable. The region has not made satisfactory efforts thus far to develop and efficiently use alternative energy sources such as natural gas, coal, and hydro. This means that the Pacific region still has room for enhancing flexibility of its energy supply and demand structure. Efforts for the development and greater use of alternative energy sources are necessary in order to improve its energy mix. While considering adequate and well-balanced allocations of energy sources, the countries of the region also must carefully examine their economic future needs.

As shown in Figure 3.1, the shares of energy trade (imports and exports) of

Asia and Oceania are higher than its shares of reserves, production, and consumption, regardless of energy resources. This means the existence of both the resource-rich and the resource-poor in the Pacific region, and suggests that energy trade has a key role in producing and consuming countries alike. In this sense, it is essential for the Pacific region to develop a healthy energy trade, resting on free trade, and to establish an open and attractive market.

ENERGY DEMAND

The past and expected future trends of primary energy demand in the Pacific will be briefly surveyed here. Projected figures for 1990 and 2000 are taken from forecasts made by individual governments or research institutions in the Pacific region. For a few countries for which forecasts are not available, the growth rates were estimated from their past trends.

Primary energy demand, totaling 56.9 million barrels per day (b/d) (oil equivalent) in 1975, grew at an annual rate of 2.6 percent and reached 64.7 million b/d in 1980. Then, slowing down because of the influences of energy conservation and economic recession resulting from the second oil crisis, it remained at 65.8 million b/d in 1984, up 0.4 percent from the previous year. Above all, oil demand has been sharply declining since the second oil crisis. As a result of the growing use of alternative energy (coal, natural gas, hydro, geothermal, and nuclear), oil dependence dropped from 48 percent in 1975 to 41 percent in 1984.

Growing at an annual rate of 2.5 percent, primary energy demand in the Pacific region is expected to reach 97.7 million b/d in 2000. This figure appears to be rather high because it was obtained simply by adding up the targets planned by individual governments. Many recently announced forecasts for the world energy supply and demand estimate the growth of primary energy demand at around 2 percent a year. Hence, the Pacific region is expected to register a relatively high growth of primary energy demand. Although oil demand will be on the rise in absolute terms, from 27.3 million b/d in 1984 to 33.5 million b/d in 2000, oil dependence is likely to fall to 34 percent, reflecting further progress in alternative energy sources. The share of energy consumption by developed countries is projected to drop from a high of 81 percent in 1984 to 68 percent in 2000. However, this means that the share of energy consumption by developing countries will rise by an equivalent amount and will account, to a large extent, for the expected sharp growth of primary energy demand in the Pacific region.

In this connection, primary energy demand in developing countries is predicted to grow at an annual rate of 5.7 percent until 2000. Such a high growth of energy demand and the ambitious goals this implies for alternative energy plans depends on whether these countries can attain favorable economic development. With international division of industries and labor expected to be in progress among developing countries, newly industrializing countries (NICs), and developed countries, it is probable that the Pacific region will undergo drastic changes in industrial structure. This means that the shifting industrial structure will form an additional crucial factor in the growth of energy demand in the developing countries of the region.

Local Characteristics of Primary Energy Demand

As shown in Figure 3.2, major areas in the Pacific region registered marked drops in their oil dependence between 1975 and 1984. Nevertheless, the current oil dependence of the Pacific region still outpaces the world average (around 38 percent). The downward trend of oil dependence is likely to continue to 2000. Particularly sharp declines are expected in the ASEAN and East Asian countries.

Progress in alternative energy introduction has been a primary contributor to the reduction in oil dependence between 1975 and 1984. Judging from the growth rates of primary energy demand in addition to the data on oil dependence, the alternative energy sources that are registering marked gains are natural gas, coal, hydro, and geothermal in Southeast Asia; nuclear and coal in East Asia; natural gas, nuclear, and coal in Japan; natural gas in Oceania; coal and nuclear in the United States; and coal in the centrally planned economies (CPEs). The rise of these alternative energy sources resulted from strong efforts by individual countries to diversify supply sources and promote more effective energy use. Countries with available, untapped domestic energy sources endeavored to develop them, and those without domestic energy potential looked to imported alternative energy sources.

Until 2000, countries in the region intend to pursue alternative energy plans, regardless of whether this involves imported resources. Strategies differ across the subregions. The ASEAN members intend to expand the use of coal, natural gas, and hydro; East Asia, liquefied natural gas (LNG), nuclear, and coal; Japan, nuclear and LNG; Oceania, natural gas and coal; North America, nuclear and coal; and the CPEs, coal and hydro. The ASEAN members, East Asia, and the CPEs are expected to have high growth rates of primary energy demand, reaching 7.4, 5.4, and 5.4 percent, respectively; sharp increases are also likely in their use of indigenous and imported alternative energy resources. In the years leading to the twenty-first century, these areas are expected to play increasingly important roles in raising energy consumption in the Pacific region.

As mentioned, some developing Pacific countries, primarily East Asian states like the Republic of Korea and China, have already started nuclear power generation or initiated construction of nuclear power stations. In parallel with their economic development in the years ahead, these countries are expected to show growing ratios of nuclear power generation. Also, some of the other countries in the region, including Indonesia, are now giving serious consideration to the promotion of nuclear power generation in the future. Therefore, to foresee the Pacific energy supply and demand mix for 2000, the moves in these countries deserve to be watched.

Figure 3.2
Historical Trends and Prospects of Primary Energy Demand Structure in the Pacific Region (percent of oil dependency)

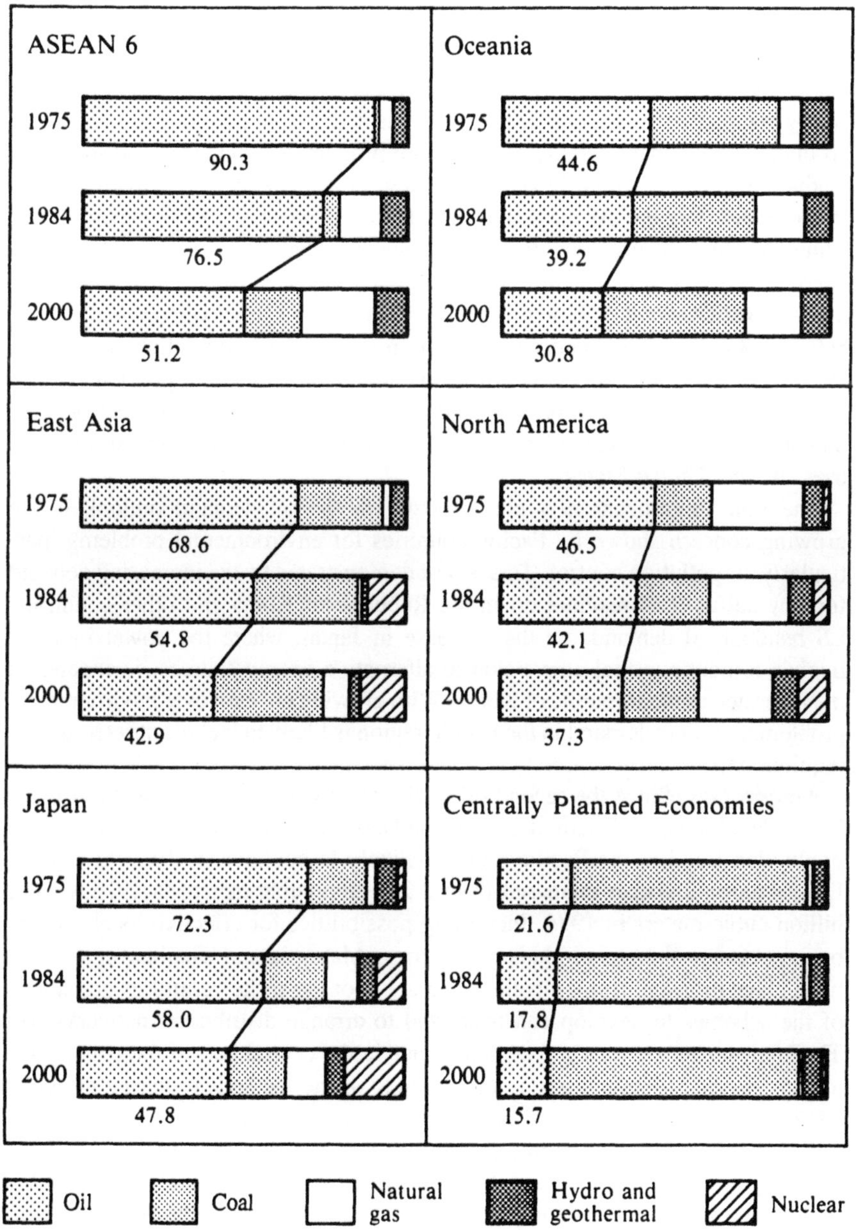

Characteristics and Problems by Primary Commercial Energy Source

Oil. Between 1975 and 1984, efforts were made to reduce oil dependence, the most effective being a sharp decline in industrial residual oil demand attributable to replacement by alternative sources of energy. However, partly because viable substitutes were absent, demand for gasoline and intermediate distillates kept growing at a relatively steady pace. As a result, a greater share of lighter products is now characteristic of petroleum product demand. Demand for intermediate distillates, primarily kerosene and gas oil, is likely to continue growing in the Pacific region. At the same time, there are signs of continuously diminishing residual oil use as a result of increasing popularity of such alternative energy sources as coal, natural gas, hydro, and nuclear. Hence, it seems likely that the shift toward lighter products will be further intensified. Wider gaps between supply and demand stemming from greater demand for lighter products are foreseen for individual countries, potentially causing grave problems.

If oil prices continue to be low, demand switching from coal to fuel oil may increase. This switching tendency would become especially conspicuous in countries like the United States.

The other significant new characteristic in the oil demand outlook is the growing concern shown by Pacific countries for environmental problems, particularly air pollution control. This is best demonstrated by the increasing demand for low sulfur (LS) residual oil in the Republic of Korea and other countries. LS residual oil demand, on the decrease in Japan, where the power industry intends to continue fuel conversion to alternative energy sources, is expanding in NICs such as the Republic of Korea. With growing concern over environmental problems, product demand in the Pacific region is likely to include stricter quality requirements.

Natural Gas. Since the second oil crisis, remarkable progress has been made in the development of natural gas in Southeast Asia and Oceania. The upward production trend in the Pacific region is likely to continue in the years ahead. LNG exports to Japan increased from 2.2 billion cubic meters in 1975 to 30.2 billion cubic meters in 1984. Increasing possibilities for effective local use can be expected in the gas-producing countries. Malaysia and Thailand reportedly have plans to install trunk pipelines and are about to make careful examinations of the schemes to develop end users and to arrange distribution networks. (In Thailand, power stations and cement plants in the central part of the country are already fueled by natural gas.) Further study of the methods to permit effective gas consumption in individual producing countries is needed.

Lately, liquefied petroleum gas (LPG) use in developing countries has been growing rapidly because LPG replaces kerosene in the commercial and residential sectors and is a substitute for gasoline. LPG consumption in the ASEAN countries, which was 370,000 tons in 1975 and 590,000 tons in 1980, increased to 1.05 million tons in 1984. Although LPG use had been restricted because of safety problems, these countries are striving for greater LPG use because its output is growing as a result of increasing capacity, including natural gas processing plants, associated gas recovery plants, and petroleum gas recovery plants in the refinery sector. These moves seem likely to become more conspicuous in coming years. It will become crucial to give consumers adequate education and guidance on safety management.

Coal. Coal demand in the Pacific region is rapidly growing due to fuel conversion in the cement industry from residual oil to steaming coal, as well as construction of new coal-fired thermal power stations. A time-series review of the changes in the coal demand in 1975, 1980, and 1984 reveals that coal demand started expanding in Japan (81, 87, 105 million tons), followed by the Republic of Korea (17, 25, 33 million tons), Taiwan (3.4, 6, 10 million tons), Hong Kong (0, 0, 4 million tons), the Philippines (110,000, 580,000, 1.6 million tons), Thailand (0, 10,000, 140,000 tons), and Malaysia (20,000, 50,000, 300,000 tons). This may be one of the best examples of an industrial activity spreading from the developed countries to the NICs and to the developing areas. Because the cement industry has already completed its coal conversion to a considerable extent, incremental coal demand in the cement industry, if any, will be attributable not to oil-to-coal conversion but to additional coal needs resulting from new cement demand. The installation of new coal-fired power generating capacity is likely to act as the prime mover in future growth of steam coal demand. Judging from present conditions, it appears difficult to boost steam coal use in any other field.

With a new steel mill coming onstream, coking coal demand is growing in the Republic of Korea. The coal consumption in the iron and steel industry increased from 849,000 tons in 1975 to 6.89 million tons in 1984. Although little growth of coking coal demand can be expected in advanced countries, due to their sluggish steel production, the NICs (particularly the Republic of Korea) are likely to experience continued growth of their crude steel production and to show growing coking coal demand. Coking coal demand as a whole will increase slightly because of offsetting trends in demand in the developed nations (down) and in the NICs (up).

Some developing countries, such as Thailand and Indonesia, have deposits of low-grade brown coal. Effective use of such indigenous resources has become a high-priority project for them, partly because they hope to save foreign currency by substituting these for imported energy. Production of briquettes or methanol from brown coal will be a viable means of effective utilization. The Philippines is considering the use of domestic coal in power generation. But to follow this line forward, the country must solve knotty problems in utilization technology, including the need for blending with higher-grade overseas coal.

In China, coal is the most important energy source; its consumption in 1984 amounted to 750 million tons, dominant among coal consumers in the world. With the production target set at 1.2 billion tons by 2000, most of which will be for domestic consumption, China intends to continue its efforts in expanding domestic coal use. Insufficient transport capacity combined with maldistribution of resources forms one of the key factors in China's serious energy problems.

One of the characteristics of coal use is the likelihood of environmental problems, exemplified by acid rain. Although Pacific countries are now becoming more and more conscious of environmental problems, further promotion of effective coal use will require more consideration be given to such problems.

Breakaway from Noncommercial Energy Use by Developing Countries

Apart from the commercial energy sources discussed so far, noncommercial energy sources like firewood, charcoal, and biomass are in use in developing countries. Few statistics can show the exact size of noncommercial energy demand, though the United Nations' *Energy Balance and Electricity Profiles 1984* suggests that the share of noncommercial energy in total energy demand in Southeast Asian developing countries (excluding Singapore) ranges from a low of 10 percent to a high of approximately 50 percent.

Soaring oil prices, triggered by the second oil crisis, pushed up the price of kerosene used for cooking in developing countries. On the other hand, subsidies to help the public cover fuel costs were cut in these countries, including Indonesia, after oil revenues fell with world prices. Consequently, the domestic price of kerosene for cooking rose. The Indonesian government repeatedly increased local kerosene prices between 1982 and 1985 (37.5 rupiah before January 1982, 150 rupiah from January 1984, and 165 rupiah from April 1985). As a result, there may be a shift back to noncommercial sources such as firewood and agricultural wastes in rural areas. Thailand, Indonesia, and Malaysia have been estimated to have average annual noncommercial energy growth rates of 3.0, 2.1, and 2.8 percent, respectively.

The use of noncommercial energy can be evaluated from the viewpoint of planned energy diversification. In the absence of any plans, expanded noncommercial energy use may devastate valuable forest resources and cause environmental problems. Improvements in living standards, health, and convenience require a suitable energy mix, with increasing importance given to commercial energy. Apart from kerosene, potential cooking energy sources include briquettes manufactured from brown coal and LPG. Though falling in the category of noncommercial energies, charcoal of improved quality and higher calorific value can also be viable. At any rate, the crucial problem is whether developing countries, taking the living standards and health conditions of consumers into full consideration, can supply adequate energy products.

Electrification in Developing Countries

As pointed out by Professor Arismunandar, director general of electricity and new energy, Department of Mines and Energy, Indonesia, at the International

Symposium on Energy held on June 6, 1986, electrification in developing countries is motivated by an unconditional, ardent wish among the public for improving living standards by making electric lighting available. There are positive interrelations between per capita GNP and per capita power consumption in developed countries, NICs, and developing countries in the Pacific region. In the period 1976–1984, power consumption grew in NICs and developing countries alike.

Electrification is significant for several reasons, including the following:

- Progress in industrialization backed by upgraded power supply capacity
- Improved national life, resulting from better availability of power supply
- Innovations of social forms and structures in the medium and long term, based on accumulated effects of electrification promotion.

Electrification can be promoted both by large-scale, centralized power generation and by decentralized power generation. The former involves large-scale power resource development as well as installation of transmission and distribution networks, and is mainly designed to supply electric power required for industrialization and for commercial and residential use in and around big cities. The latter represents localized power generation, which is of high potential as a means to facilitate electrification in rural areas. In developing countries, centralized and decentralized generating systems successfully coexist, each playing its own role. The typical example is seen in urban areas, where in-plant generation provided by diesel power generation plays an important role.

Once electrification is in progress and transmission and distribution networks are installed to some extent, various problems, including fluctuating voltage and blackouts, can emerge. Factors such as insufficient mechanical maintenance, unskilled operating techniques, and overloaded access can result in a serious situation, preventing electric power from reaching potential users. Some developing countries are now suffering from serious power shortages, indicating that many of the countries committed to electrification without giving full consideration to those problems until they emerged.

In the course of electrification, developed countries and NICs have enjoyed a chain of positive effects:

- The development of power resources, originally aiming at the security of lighting power sources, provides additional generating capacity
- Industry is allowed to use this extra electric power
- Industrial development creates additional power demand
- Additional electric power stemming from new power resource development encourages installation of additional transmission and distribution networks
- Expansion of the networks leads to improved national life.

Moreover, the costs incurred in these processes have been offset by improved income levels resulting from industrial progress. Thus, for successful promotion of large-scale centralized power generation, a good combination of progress in industrialization and improvement of national life is required.

In regard to electrification of rural areas through decentralized power generation, it must first be determined if living standards in a given area can justify introduction of a given generating system in economic terms, and if the local staff is competent for system management in terms of both operation and maintenance. If the answers to these questions are positive, decentralized power generation could lead to the installation of end-user networks, ultimately permitting interconnection with trunk networks.

Meanwhile, the promotion of electrification, first called for by Professor Kojima of Hitotsubashi University at the Symposium on Energy and Development in Asia, jointly sponsored by the Ministry of Foreign Affairs, Japan, and the Keidanren (Federation of Economic Organizations) in February 1985, was advocated anew by Mr. Mizoguchi, ambassador and special adviser to the minister for foreign affairs, in his presentation of a concept for "the second electrification drive" at the Symposium on Pacific Energy Cooperation, cosponsored by the Ministry of Foreign Affairs, Japan; the Ministry of International Trade and Industry; and two other energy-related private bodies in late March 1986. In July 1986, when the Minerals and Energy Forum, a meeting of the Pacific Economic Cooperation Council, was held in Jakarta, the promotion of electrification was further discussed. Thus, discussions about electrification in developing Asian countries have gradually unfolded.

ENERGY PRODUCTION AND RESOURCES

Increasing Energy Production and Supply-Demand Imbalances

Oil. Crude production in the Pacific region grew sharply between 1975 and 1984, with large output increases recorded by Malaysia, China, Australia, and Mexico. Apparent crude production in Asia and Oceania has leveled off because Indonesia complied with the production reduction and sharing policy of OPEC. As discussed later, Indonesia, Malaysia, and Australia are not necessarily blessed with proven oil reserves (19, 20, and 7 years, respectively). Without satisfactory exploration and development in the coming years, these oil producers may face shrinking crude output in the 1990s. According to a forecast by the National Petroleum Council, U.S. crude oil production is expected to go down by 2.5 million b/d, from 8.9 million b/d in 1985 to 6.4 million b/d in 2000. China set its target of crude oil production at 3 million b/d for 1990 and 4 million b/d for 2000.

One of the characteristics of refining capacity in the Pacific region is its poor cracking capacity. Considerable amounts of cracking capacity must be installed in the years ahead in order to cope with greater demand for lighter products, as already mentioned, through capacity upgrading. Some countries, such as Thailand and Malaysia, lack domestic refining capacity, while others, including Japan, Singapore, and the Philippines, are confronted with the problem of excess topping capacity. If these countries intend to count on domestic refining alone, they must face the need for huge capital investment to install topping and cracking capacities. Construction of domestic refineries and installation of cracking capacity are perhaps among the national priorities of China, where sharp growth of petroleum product demand is very likely.

Natural Gas. Asia and Oceania registered a markedly higher growth rate of natural gas production in the 1974–1984 period than the rest of the Pacific region. Natural gas production in China has been limited. Though huge in volume, natural gas production in North America has been on the decline. In South Asian countries, gas production is very likely to keep growing. To put these natural gas resources to effective use, it is essential that gas producers endeavor to expand domestic gas use. Research and development efforts for new LNG applications would help toward this end.

Coal. Coal production in the Pacific region has kept growing, primarily in the four major producing countries—Australia, Canada, the United States, and China. Because coal resources are abundant, there would be little fear of a supply shortage if production levels are maintained at the demand level. Actually, there are signs of overproduction, stemming from the active development efforts initiated after the second oil crisis in hopes of sharp rises in coal demand as an alternative energy source in the long term. This forms one of the problems for the present.

Assessment of Energy Resources and Their Development Potential

The potential for future development of energy resources in the Pacific region is illustrated in Figure 3.3. Judging from its proven crude oil reserves, the number of recoverable reserve years at current rates of drawdown in the Pacific region is 18, much shorter than the world average of 33 years. Judging from ultimate recoverable reserves, the number of crude oil recoverable years in the Pacific region is 46 years. As for natural gas, judging from proven reserves, the number of recoverable years is as short as 17 in North America but more than 50 in Asia and Oceania, including the CPEs, virtually the same as the world average. Judging from ultimate reserves, the number of recoverable years of natural gas appears to be much greater than that of crude oil. Although information is available only on recoverable reserves for coal, the number of recoverable years (155) of coal in the Pacific region is much greater than that of crude oil and natural gas. Moreover, this figure covers bituminous coal only. With low-grade coal such as lignite taken into account, the number of recoverable years of coal should be even longer.

Thus, compared with natural gas and coal, crude oil resources in the Pacific region are apparently limited. Nonetheless, because ultimate recoverable reserves suggest that it is still possible to increase crude oil output within the region, further efforts for exploration and development must be made. In regard to the more abundant natural gas and coal resources, it seems essential to promote their

Figure 3.3
Possibilities of Energy Resource Development in the Pacific Region

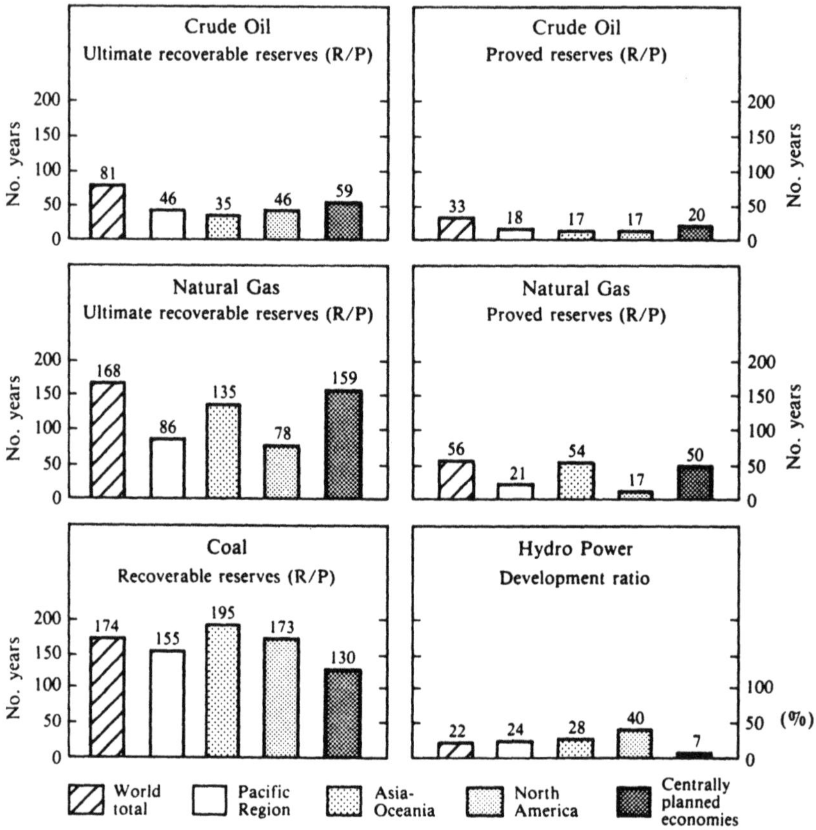

Note: R/P = recoverable years (reserve/production).

development on the basis of the concept of energy mix, stressing adequate and well-balanced allocations of energy sources in the medium and long term on both supply and demand sides.

As indicated in Figure 3.3, the development ratio of hydro resources is as high as 40 percent in North America. On the other hand, with the development ratios still low in Asian developing countries and China, hydro resources can be positioned as promising indigenous alternative energy resources in these countries.

ENERGY TRADE

Crude Oil Trade and Dependence on the Middle East

Though some changes in Pacific crude oil trade have occurred since 1974, a major characteristic of crude oil trade in the region is its heavy dependence on the Middle East. This shows that decentralization or diversification of supply sources is difficult to attain. However, comparing the trade flows of 1975 and 1984 reveals that the reduction of North American and Japanese crude oil imports has steadily pushed down crude imports from the Middle East through diminishing oil demand. The reliance on the Middle East in securing crude oil imports dropped from 60 percent in 1975 to 49 percent in 1984. (With North American imports excluded, the dependence on the Middle East stood at 81 percent in 1975 and 68 percent in 1984).

The Pacific region has no oil supply sources as large as those of the Middle East. However, as a result of the two oil crises, constant efforts have been made to develop oil resources. This has led to output increases by Malaysia, China, Australia, Alaska, and Mexico, whose export capacities have been increasing. Looking at 1984 trade flows, in East Asia and Southeast Asia, where crude imports are growing, increasing amounts of crude oil are being imported from such intra-area sources as Indonesia and Malaysia. As a result of processing deals with China, Singapore is refining growing amounts of Chinese crude. Australia reduced its imports because of higher domestic crude production. Crude exports by China and Mexico are steadily rising. Thus, although diversification of supply sources is not easily achieved in the Pacific region, crude oil trade within the region increased between 1975 and 1984.

However, considering the increasing oil demand and decreasing crude production in Malaysia, Indonesia, and Australia, it is highly probable that their crude oil export capacity will become lower if its oil demand grows at a higher pace than projected by the Chinese government. Perhaps Mexico will be the only producer in the region still able to maintain its export capacity in 2000. The U.S. Department of Energy estimates that the country's net oil imports, totaling 4.7 million b/d in 1985, will grow to 8 million b/d in 2000 and to 9.4 million b/d in 2010. The incremental needs are so large that growing imports from the Middle East should be inevitable, even if they could be covered by imports from Central and South America and Africa at the initial stage. Taking everything into account, the dependence on the Middle East for securing crude imports is quite likely to resume its rise, thus posing a crucial issue for the future.

Expansion of Product Trade and Elimination of Supply-Demand Imbalances

Some major changes have occurred in petroleum product trade in the Pacific region. The share of product imports in total petroleum imports, including crude oil imports, grew from 20 percent in 1975 to 25 percent in 1984 (13 percent and 17 percent, respectively, with North America excluded), thus demonstrating steady expansion of product trade. Looking at 1975, product imports of 1.81 million b/d by North America alone dominated. In 1984, however, product imports were also growing in the ASEAN countries, from 260,000 b/d in 1975 to 370,000 b/d in 1984; East Asia, 150,000 b/d in 1975 to 190,000 b/d in 1984; and Japan, 250,000 b/d in 1975 to 490,000 b/d in 1984.

Focusing on product exports, Singapore increased its refining capacity from 50,000 b/d in 1965 to 920,000 b/d in 1975 and to 1.1 million b/d in 1985, and boosted product exports from 360,000 b/d in 1975 to 590,000 b/d in 1984, thus providing a balancing function in Asia. While boosting its crude production, China also increased exports of gasoline, naphtha, and gas oil, from 14,000 b/d in 1975 to 131,000 b/d in 1984. By decontrolling product exports in October 1981, the United States also intensified its product export drive to the Pacific region. The United States put special emphasis on exports of residual oil, which has been in excess supply at home because of diminishing demand. Although U.S. petroleum product exports were minor in 1975, they increased to 110,000 b/d in 1984. Though not so large in quantity, all East Asian and ASEAN countries engage in product exports as a means to adjust local product supply and demand.

The reliance on the Middle East for securing product imports has not been as high as for crude imports. In 1984, it stood at 8 percent. Even excluding North America, it was as low as 17 percent. However, considering that producers in the Middle East, mainly Saudi Arabia and Kuwait, have made progress over the years in constructing new and additional refineries, the reliance on the Middle East for product imports is likely to rise gradually in parallel with an increasing number of refineries coming onstream in the Middle East.

As already discussed, the Pacific region is expected to have a growing demand for oil, primarily intermediate distillates, thus showing an intense inclination toward lighter products. Some Pacific countries are in need of topping capacity and are unable to meet local needs with existing cracking capacity. Of the two options to meet growing product demand—capacity expansion and upgrading or product imports and exports—many Pacific countries view the installation of costly cracking capacity only as a long-term option, and intend to employ product imports and exports for the time being. Combined with ballooning product exports from the Middle East, product trade in the Pacific region is very likely to continue expanding in the years ahead. Amid swelling product trade, how to work out a balancing function to adjust product trade imbalances among various countries will become an important problem.

Expansion of LNG Trade

There was an upward trend in natural gas trade between 1975 and 1984 in the Pacific region. Japan and the United States are LNG importers. LNG imported by the United States comes from Algeria and is limited in quantity. In 1975, Japan's LNG imports were still small in size, with Alaska and Brunei being the only two sources of supply. Between then and 1984, however, projects to develop natural gas resources were carried out by Abu Dhabi in the Middle East as well as Indonesia and Malaysia in Southeast Asia. These enabled Japan to boost its LNG imports.

Japan's LNG import plans include 5.84 million tons of Australian gas to be received in 1989. Until recently, Japan had been virtually the only LNG importer in the Pacific region, but in 1986 the Republic of Korea began to import it. Korea's LNG imports (supplied by Indonesia) are expected to reach 2 million tons in 1987 and to rise to 5 million tons a year in 1996. Taiwan also plans to initiate LNG imports of 1.5 million tons from Indonesia in 1990. A few more LNG projects planned for 1990 and afterward are under consideration in the Pacific region.

Natural gas exports to Singapore, but not in the form of LNG, are planned for 1990 and afterward upon the completion of the Trans-Malaysia Pipeline. Natural gas trade through pipelines has been popular in North America for many years.

Compared with other energy trade, present LNG trade is simple, with little progress being made in diversification of destinations. Despite such a trade climate, natural gas production and development in the Pacific region and the rest of the world are booming. Hence, more is required than merely encouraging effective domestic use by gas producers themselves; one must also examine the potential of new LNG importers. Conventional trade terms, such as the take-or-pay clause, have functioned well enough to facilitate giant development projects and to stabilize production. To stimulate LNG trade in the future, however, more flexible trade terms seem crucial. This would entail the introduction of a new pricing system linked with actual energy market conditions, and review of the take-or-pay clause and other provisions causing present supply rigidity. Faced with decreases in crude oil prices, gas producers have been showing more and more flexibility in pricing.

Expansion of LPG Trade and Diversification of Supply Sources

Recently, LPG consumption in the Pacific region has been rising sharply in developing countries, and LPG trade has concurrently become active. Because LPG imports by the Pacific countries accounted for 68.5 percent of the world total in 1984, LPG trade holds an important position in energy trade. It also was on the increase in the Pacific region between 1975 and 1984. Apart from Japan, LPG imports were increased by the Republic of Korea, Taiwan, Hong Kong, and Thailand. Japan's imports doubled between 1975 and 1984. East Asian and ASEAN countries, where LPG imports were virtually absent in 1975, imported some in 1984. LPG trade also expanded in North America, which is largely attributable to the growth of trade within the area.

The Middle East has been the primary supplier to cover growing LPG trade in the Pacific region. The dependence on the Middle East in securing LPG imports jumped from 44 percent in 1978 to 54 percent in 1984 (72 percent and 81 percent, respectively, with North America excluded). Though similar to crude oil imports in the aspect of heavy reliance on the Middle East, LPG imports differ from crude imports in that they increased rapidly up to 1984. This increased the dependence on the Middle East. At present, LPG supply sources are heavily concentrated in the Middle East, where LPG supply sources fluctuate, depending on crude production levels. This poses major problems. LPG from Singaporean refineries is shipped largely to the ASEAN countries and Hong Kong.

Lately, some Pacific countries have made active moves to install LPG recovery facilities at local natural gas processing plants, associated gas processing plants, and refineries. For instance, Indonesia plans to install a facility to separate and recover LPG at an LNG plant and produce large amounts of LPG for export. Similar moves of LPG separation and recovery are growing in Malaysia, Thailand, Australia, and New Zealand. While endeavoring to expand domestic LPG use, these countries are also expected to contribute to the growth of LPG trade. Diversification of supply sources in the Pacific region, backed by LPG output increases, can help lower the region's reliance on the Middle East in securing LPG trade.

Expansion of Coal Trade and Diversification of Destinations

A remarkable growth of the Pacific coal trade occurred between 1975 and 1984. As in natural gas trade, Japan was virtually the only coal importer in the region in 1975. (In North America, coal was imported mostly from the intra-area countries.) However, in 1984, when the cement industry committed to coal conversion and large numbers of coal-fired thermal power stations were put onstream, steam coal imports by East Asian NICs and ASEAN countries, in addition to Japan, grew sharply, thus diversifying destinations of coal shipments. Increasing coking coal imports by the Republic of Korea also increased Pacific coal trade.

Primarily backed by strong demand from the power generation sector, steam coal imports by Japan, the NICs, and the ASEAN countries are likely to keep growing. Results of a survey on expansion of the Pacific coal flow, conducted by the Institute of Energy Economics of Japan, show that steam coal imports by these countries are likely to total around 65 million tons in 1990 and to exceed 100 million tons in 1995, compared with some 35 million tons registered in 1984. As for coking coal, the NICs (mainly the Republic of Korea) are projected to raise their imports in parallel with rising crude steel output. Sluggish steel production levels are expected in developed countries; consequently, little growth of coking coal imports can be expected on the whole.

There are five major coal suppliers: Australia, Canada, the United States, South Africa, and China. In addition, with studies on coal development under way in East Kalimantan, Indonesia has been gradually strengthening its position as a coal exporter in recent years. In 1984, for example, shipments of Indonesian coal to Japan, the Republic of Korea, Taiwan, and Malaysia were 300,000, 60,000, 60,000, and 330,000 tons, respectively. Coal exports from Indonesia were virtually absent in 1975.

As with other energy resources, interests and claims of coal producers and consumers have generally been "passing each other." So far, adjustment of the ratio of spot to long-term contracts has been effective to some extent. However, recent experiences have revealed that the exact fulfillment of contract terms may abruptly become impossible because of sharp fluctuations in exchange rates, steep ups and downs in competitive energy price levels, and changing industrial structures in consuming countries, all of which make immediate demand and supply adjustments difficult. This problem requires both producing and consuming countries to take adequate measures by giving full consideration to the lead time between the development and availability of coal resources.

As outlined, regional energy trade expanded remarkably between 1975 and 1984. The upward trend appears likely to continue in the years ahead, involving all of the developed countries, NICs, and developing countries in the region. The Pacific region consists of three continents facing each other across the Pacific Ocean and a large number of islands. Some of the Pacific countries are rich in resources; others are intensive consumers. Thus, energy trade based primarily on marine transport can be regarded as a vital activity, indispensable for the future of the Pacific region. Stepped-up efforts must be made for smooth energy trade flows and for solution of the problems that have been pointed out.

NOTE

1. Asia and Oceania as described in this chapter is the region, excluding North America and Asian centrally planned economies of the Pacific region, defined as including ASEAN, East Asian NICs, Australia, New Zealand, and the Pacific Islands.

FOUR

Changes in the Balance of Payments and Capital Flows: Implications for Asia

WILLIAM E. JAMES

INTRODUCTION

The rapid real economic growth that occurred in East and Southeast Asian developing countries in the 1970s surprised many observers. The 1960s are now regarded as a golden age of stable, steady expansion of world economic activity, especially compared with the 1970s. The United States enjoyed great prosperity and Japan had double-digit growth rates.

But the early 1970s saw high inflation and the collapse of the Bretton Woods system, both of which preceded the first oil shock in 1973–1974. A steep worldwide recession ensued as a result of that oil shock. Growth was sharply down in the United States and Europe. Japan had largely completed its structural transformation from agriculture to industry, and its growth rate was halved. By the mid–1970s stagflation had become the nemesis of developed and developing countries alike.

The oil shock led to a radical shift in the current account situation as a number of Middle East OPEC countries piled up huge surpluses. The recycling of these petrodollars to deficit countries (mainly LDCs) was rapidly achieved, but resulted in huge debt obligations in many deficit countries (Campbell 1982).

In spite of the recession, real economic growth in East and Southeast Asia continued. Within the region, the oil-importing developing countries had growth rates as high as or higher than those of the 1960s. Oil exporters like Indonesia and Malaysia experienced new prosperity as budgetary and foreign exchange constraints seemed nonexistent.

In the mid–1980s stark changes occurred in the world economy and in the growth performance of most East and Southeast Asian developing countries. Economic growth was far below the near double-digit rates of the 1960s and

Table 4.1
Growth of Real GDP and GDP per Capita (compounded rates)

Country	Real GDP					Real GDP per capita
	1960–70	1970–80	1980–85	1986	1987	1960–85
NICs						
Hong Kong	10.1	9.8	4.2	11.0	12.0	6.3
Korea	9.5	8.2	7.5	11.9	11.2	6.4
Singapore	9.2	9.3	6.0	1.8	8.6	6.7
Taiwan	9.6	9.7	6.1	10.6	11.2	6.4
ASEAN-4						
Indonesia	3.8	8.0	3.6	3.2	3.5	3.0
Malaysia	na	7.9	5.1	1.2	2.0	4.2[a]
Philippines[b]	5.2	6.3	-0.9	1.1	4.6	1.6
Thailand	7.9	6.9	5.3	3.5	6.6	4.2
South Asia						
Bangladesh	na	5.8[c]	3.9	4.4	4.5	2.5[d]
Burma	2.8	4.2	5.2	3.5	3.0	1.8
India	3.9	3.2	5.4	5.0	1.5	1.6
Nepal	2.2	2.0	3.9	4.2	2.0	0.1
Pakistan	na	5.6[c]	6.3	7.8	7.0	2.7[d]
Sri Lanka	5.8	4.7	5.0	4.0	3.0	3.2
China, People's Republic of	4.0	5.8	9.8	7.4	9.0	3.9

na = Not available.

a. 1970–85. b. GNP. c. 1973–80. d. 1973–85.

Sources: International Monetary Fund, *International Financial Statistics*, Yearbook 1986 and monthly issue, May 1987 (Washington, D.C.: IMF, 1987); Asia Development Bank, *Annual Report 1986* (Manila: The Bank, 1987); World Bank, *World Tables 1980* (Washington, D.C.: World Bank, 1983); Republic of China, *Statistical Yearbook of the Republic of China, 1986* (Taipei: Republic of China, 1986); People's Republic of China, State Statistical Bureau, *Statistical Yearbook of China 1986* (Beijing: The Bureau, 1986).

1970s (Table 4.1). From 1980 to 1985 real growth in the newly industrializing countries (NICs) of East Asia—Hong Kong, Singapore, Taiwan, and the Republic of Korea—slowed to the 4 to 7 percent range. In 1986 and 1987, however, Taiwan, South Korea, and Hong Kong returned to a high growth path, averaging better than 10 percent increases in real output. The NICs were aided by some external factors, such as the yen appreciation and the fall in oil prices, but they also succeeded in making rapid structural changes.

The four large developing countries of the Association of Southeast Asian Nations (ASEAN–4)—Indonesia, Malaysia, Thailand, and the Philippines—faced with heavy debt burdens and declining export earnings, were forced to reduce expenditures, imports, and growth to conserve foreign exchange. From 1980 to 1985 average real growth rates in the ASEAN–4 countries fell sharply from the 1970s levels. Except for Thailand, the ASEAN–4 countries since 1984 have been economically stagnant, experiencing negative or near zero growth. Clearly, the ASEAN–4 have not coped well with the changes in the world economy in the mid–1980s.

In low-income South Asian nations the rate of economic growth has been less a function of world trade and investment conditions than of internal conditions; in particular, performance of agriculture and related agriculturally based industries. The growth pattern of these countries seems to have improved from the late 1970s to the present, with improvements in agricultural production and reforms to increase economic efficiency. Some of these countries remain acutely in need of capital inflows of a concessional nature to continue development, though India has increasingly been able to borrow on world capital markets.

In this chapter we focus on some major changes in the balance-of-payments situation of these Asian countries and their implications for the Asia-Pacific regional economy.

CHANGES IN THE BALANCE OF PAYMENTS

Since the end of the serious world recession in 1983, dramatic changes have occurred in the balance-of-payments positions of various groups of countries. From 1983 to 1987 the aggregate current account position of high-income oil-exporting countries has been in deficit. This is a remarkable shift from the huge surpluses these countries, particularly Saudi Arabia, experienced between 1974 and 1982. Among the OECD nations equally dramatic changes have occurred. The United States has experienced deficits of unparalleled size, while West Germany has been running substantial surpluses. One of the most significant changes is that a new group of capital-surplus countries appears to be emerging in East Asia.

Japan's Capital Surplus

Japan had alternating periods of current account surpluses and deficits in the 1970s, with deficits concentrated in years of oil price increases (1974, 1979, and 1980). Since 1982, Japan's current account position has been one of huge and mounting surpluses. In 1985 Japan's current account surplus reached $49.3 billion (all money in this chapter is expressed in U.S. dollars), even greater than Saudi Arabia's surpluses during the height of the oil boom in 1980 and 1981.

In 1986, the current account surplus of Japan was $88 billion. The surplus of the oil-exporting developing countries has exceeded that amount in only a single year, $104 billion in 1980. The current account surplus of Japan, Korea, and Taiwan combined in 1986 reached $105.6 billion!

Japan has become the principal external financer of the U.S. deficits and is also the predominant capital supplier in the entire Asia-Pacific region. Despite the appreciation of the yen against the dollar and most other currencies, Japan's surplus grew larger in 1986. The collapse of oil and commodity prices (about 80 percent of Japan's imports are raw materials, petroleum, and food) contributed to the surplus. The strong yen has important implications for trade and capital flows within the region and will be considered below. The sharp appreciation of the yen has also had some adverse repercussions for Japan's domestic economy. Real economic growth fell after 1985 and unemployment rose as many businesses were rendered uncompetitive. The problems associated with the strong yen have forced Japan to make some major structural changes. Many Japanese businesses are seeking to relocate facilities or to find new, less expensive sources of supply outside Japan. Despite these difficulties, Japan has been able to recover quickly; real growth increased in 1987. The size of the current account surplus seems to have peaked in 1986, and domestic demand now appears to be the driving force behind Japan's recovery.

The NICs: New Exporters of Capital for Asian Development?

Besides Japan, a new group of small capital-surplus countries outside the OECD's traditional club of "industrial countries" appears to be arising in East Asia. These are the East Asian NICs: Hong Kong, Singapore, Taiwan, and South Korea. Although Singapore is a member of ASEAN, it is considered to be a NIC because of the development of its industrial sector and its high per capita income.

In a pattern similar to that of Japan, Taiwan's modest current account surpluses in most of the 1970s and 1980s were partly offset by large deficits in the years of the two oil shocks (1974 and 1980). Since 1981, Taiwan's surpluses have increased spectacularly, reaching an estimated $13 billion in 1986. On a per capita basis, Taiwan's surplus is even larger than Japan's.

In addition to Taiwan, other NICs appear to be moving into capital surpluses as well. Hong Kong and Singapore should be distinguished from the other NICs when considering the balance-of-payments situation. They are both international financial centers, and changes in their foreign reserves are often dominated by capital flows rather than current account transactions. Nevertheless, Hong Kong has had two successive years of substantial current account surpluses (estimated at between $1.5 and $2 billion). In 1985 Hong Kong's real growth rate was less than 1 percent and the current account surplus was due to import reductions. In 1986, though, Hong Kong benefited from the strong yen as exports surged by almost 20 percent. Hong Kong's investment climate has improved considerably as Japanese and local private investors have pumped more money into manufacturing, construction, and services.

In contrast, Singapore has recovered more slowly from the recession it experienced in 1984 and 1985. Singapore had two years of subnormal economic growth, and its exports were adversely affected by slowed growth in neighboring

ASEAN–4 countries. Indonesia, which previously relied on Singapore to refine crude oil, now has its own refineries; this has also hurt Singapore's exports. Since 1983, however, Singapore's foreign currency reserves have risen sharply from $9 billion to almost $13 billion in early 1986. Budget surpluses and a fall in the amount and rate of investment, so that domestic saving roughly equaled capital formation in 1985, also tend to confirm that Singapore is heading into a capital-surplus position, even with its recent economic difficulties. Singapore adopted a number of measures to stimulate growth. In 1987 real growth recovered to about 8 percent, and a current account surplus based on strong export growth was attained. To boost investment and consumption, and reduce the costs of doing business, Singapore cut taxes and lessened the amounts that must be paid to the compulsory Central Provident Fund. With large excess capacity in major industries and in the property sector, though, it will take time for these measures to work.

None of these three economies—Taiwan, Hong Kong, and Singapore—has substantial external debt or debt service payments. They have been moving rapidly into the role of net creditors.

Perhaps the most astonishing change in Asia's balance-of-payments situation is the transformation of South Korea from a large deficit country to a country with current account surpluses of around $5 billion in 1986 and $10 billion in 1987. With outstanding foreign debt of $48 billion in 1985, Korea was regarded in some circles as the next megadebtor LDC likely to default. Instead, it has achieved a surplus position with high economic growth and low inflation.

Lower oil prices, moderation of international interest rates, and the strong yen have all contributed to the emergence of current account surpluses of increasingly larger magnitudes in the East Asian NICs. It is tempting to regard these countries as the late 1980s equivalent of the Persian Gulf sultanates, with Japan playing the role of Saudi Arabia. In reality, however, beyond the shift in the current account positions, there are few other parallels. The NICs have income levels ranging from about $2,000–3,000 in Korea and Taiwan to $6,000–7,000 in Hong Kong and Singapore. The economies of these rapidly industrializing countries are firmly based on external competitiveness, high domestic savings, private investment, and sound management. Their exports are more diversified than those of most LDCs, and far more so than those of the Arab oil-exporting nations, though they have relied greatly on the U.S. market.

Private investment has been sluggish in the mid–1980s in Taiwan, Hong Kong, and Singapore. The slower growth in private investment is partly related to domestic political uncertainties, but is also due to slower growth in demand relative to industrial capacity. Singapore's recent slow growth has been caused in part by concentration of government and private resources in a few problem industries: electronics, petroleum refining, and ship building and ship repair. As mentioned above, because of the importance of entrepôt trade, Singapore has been hurt by slower growth in ASEAN–4 countries. Hong Kong's status as a territory of China has perhaps been the major force behind the apparent excess of savings over investment there, and is probably a reflection of private investors' desire to give themselves an option to leave in 1997. External protectionist threats in major markets like the United States and Europe have also dampened investors' enthusiasm in the NICs. China's open door policy has attracted large investment flows from Hong Kong and also has increased Hong Kong's domestic exports and reexports. The direction and scope of China's economic reform program will continue to affect regional trade and investment patterns (Ebashi and Hishida 1986; James and Young 1987).

Nevertheless, the new capital surplus situation in these countries is of interest to other developing countries, particularly those with large debts that are seeking new sources of capital as a means to stimulate economic growth. The countries of Southeast Asia, which are near the NICs and already have substantial trade and investment linkages with them, are perhaps best positioned to take advantage of the NICs' new wealth.

ASEAN–4: Import Compression and Stagnation

The ASEAN–4 have been experiencing problems in rekindling economic growth after the world recession of 1982. Though they all reduced their current account deficits in 1983–1985, they did so mainly by cutting expenditures, imports, and growth. Thailand and the Philippines reported small current account surpluses in 1986, but these had quite different causes. Thailand's surplus resulted more from expanding nontraditional exports, while the Philippines situation reflected continued import reductions. Of the ASEAN–4, only Thailand has had a positive average annual rate of import growth since 1983.

In 1986 Indonesia and Malaysia had widening current account deficits related to the decline of oil prices. All of the ASEAN–4 are heavily burdened by external debt and have been hurt by sharp declines in export commodity prices and sales. The ASEAN–4 saw their export earnings fall precipitously in 1985 and 1986, after some recovery had occurred in 1984. However, declining oil prices helped Thailand and the Philippines achieve current surpluses in 1986.

The commodity bust since 1981 appears to be long-term: a result of burgeoning world agricultural surpluses, fundamental technological changes that have reduced demand for primary goods, and slower growth in the OECD countries. Economic growth in Southeast Asia appears to be stagnant except in Thailand, where growth was in the 4 to 5 percent range in 1985–1987 (Table 4.1).

The continued heavy reliance on a few primary commodities for government revenue and foreign exchange, coupled with the inward-looking and capital-intensive industrial sectors in these countries, particularly in Indonesia and the Philippines, was a disastrous combination in the mid–1980s. The heavy external debts and debt service burdens confronting these countries (Table 4.2) made it imperative that the balance of payments be stabilized, even though this meant sacrificing growth and slowing investment and structural change.

In 1987 there were some signs of recovery in the prices of major primary

Table 4.2
Total Debt and Total Debt Service of Asia-Pacific Countries, 1970, 1978, 1984

	1970[a]			1978[b]			1984[b]		
	Total debt[c]		Debt service	Total debt[c]		Debt service	Total debt		Debt service[d]
Country	US$ millions	As % of GNP	As % of exports	US$ millions	As % of GNP	As % of exports	US$ millions	As % of GNP	As % of exports
NICs									
Hong Kong[e]	2	0.1	0.0	1,253	7.2	1.3	6,580	20.7	3.0
Korea	1,972	22.4	19.4	12,729	25.4	11.6	49,172	60.6	21.7
Singapore[f]	154	8.0	0.6	1,385	17.8	2.5	3,654	20.2	2.1
Taiwan	na	na	na	3,504	13.3	6.0	9,597	16.7	6.5
ASEAN-4									
Indonesia	3,043	33.8	6.9	14,524	29.3	19.0	33,152	43.7	19.3
Malaysia[g]	390	10.0	3.6	3,055	20.5	9.7	16,770	55.7	10.6
Philippines	1,563	22.1	7.3	6,336	26.4	23.8	23,771	73.9	30.1
Thailand	727	11.1	3.4	2,652	11.6	16.0	16,674	40.1	26.4
South Asia									
Bangladesh[h]	na	na	na	2,793	32.1	14.9	5,699	40.7	17.2
Burma	117	5.4	15.9	876	18.8	17.7	2,734	42.9	44.1
India	8,050	15.1	22.0	15,773	13.3	12.5	30,533	16.7	12.8
Nepal	3	0.3	na	106	6.6	1.5	471	18.8	5.4
Pakistan	3,131	31.2	23.7	7,759	40.7	24.2	12,399	36.9	27.4
Sri Lanka	400	20.4	10.8	1,038	38.2	9.3	3,374	57.4	17.1

na = Not available.

a. Data from World Bank.

b. Data from OECD.

c. Excludes short-term debt.

d. Excludes services of IMF credit for OECD data.

e. 1983 data for service/export ratio. Also, total debt figures exclude private nonguaranteed long-term debt.

f. Total debt figures exclude private nonguaranteed long-term debt for all years and short-term debt for 1985.

g. Total debt figures exclude private nonguaranteed long-term debt and short-term debt for all years.

h. 1975 data. Debt figures exclude short-term debt.

Sources: Asian Development Bank, *Key Indicators of Developing Countries* 17 (July 1986); Organisation for Economic Co-operation and Development, *External Debt of Developing Countries, 1983 Survey* (Paris: OECD, 1984); Organisation for Economic Co-operation and Development, *Financing and External Debt of Developing Countries, 1985 Survey* (Paris: OECD, 1986); World Bank, *World Debt Tables 1985/86*, including first and second supplements (Washington, D.C.: World Bank, 1986), *World Development Report 1986* (Washington, D.C.: World Bank, 1986), *World Tables*, 3d ed. (Washington, D.C.: World Bank, 1983).

commodities that the ASEAN–4 rely on for the bulk of export earnings, such as petroleum, palm oil, coconut products, and rubber. Indonesia was able to sharply reduce its current account deficit in 1987 through some recovery of oil prices as well as expansion of nonoil exports. While it is likely that the primary sector will continue to be the major source of foreign exchange earnings for these natural-resource-based economies, diversification of exports has become the new watchword.

Thailand has been the most successful at diversifying its exports, in both the primary and manufactured goods sectors (Naya 1987). It is the only ASEAN–4 country to have expanded exports in value terms since 1984. Malaysia also has diversified, but not enough in manufactures to offset declining commodity earnings. Indonesia has started to encourage manufactured exports, but suffers from a high-cost industrial base built up under heavy protectionist policies. The Philippines has failed to move quickly enough in developing new manufactured exports and has been unable to take the bold policy steps necessary to invigorate its productive sectors.

Thailand and the Philippines exported laborers in large numbers to the Middle East in the late 1970s and early 1980s. Remittances from these overseas workers provided substantial balance-of-payments support. With the weakened outlook for oil, the earnings from this source have fallen off, and cannot be expected to rise substantially in the near future.

South Asia

The South Asian countries, with the exception of Sri Lanka, have much lower dependence on imported oil in terms of total commercial energy consumption than do the oil-importing countries of East and Southeast Asia. External trade is generally smaller in relation to GDP than in the rest of Asia. Trade orientation of some South Asian countries has undergone a subtle change since the mid-to-late 1970s. India, for example, traditionally relied on strict rationing of foreign exchange and a stringent system of import controls to maintain a sustainable external balance. Trade deficits arise as a result of the net inflows of capital, mainly from foreign governments and agencies supporting development programs. However, following the first oil shock, a new source of balance-of-payments support arose as labor remittances from Middle East countries became substantial; in the case of Pakistan, such remittances exceeded merchandise exports. Pakistan was also able to expand exports of goods to the booming Middle Eastern economies. OPEC provided generous assistance to a number of South Asian countries during the period 1976–1982.

The higher cost of oil, of course, had a negative impact on the balance of payments of the South Asian countries. But these effects were greatly offset by the above-mentioned remittances and other external financial support. The achievement of improved agricultural production led to sharp declines in the need to import grain (which for some of the countries was as important as oil).

This also freed up foreign exchange. The attitudes and policies concerning external trade were modified, and greater efficiency and growth resulted. Notable liberalization occurred in Sri Lanka after 1977, as well as in India and Pakistan, though to a lesser extent. Because South Asia relied primarily on concessional foreign capital inflows, its debt servicing has remained relatively modest. India, which made large IMF borrowings in the early 1980s to help itself adjust to the second oil shock, could face some difficulties in meeting the repayment schedule unless it can expand exports or otherwise earn additional foreign exchange. The growth performance of South Asia was slower than that of the rest of Asia in the 1960s and in most of the 1970s. But in the 1980s real growth of South Asian countries compares favorably with that of other countries, particularly with the ASEAN countries.

DEBT AND CAPITAL FLOWS

In the early 1980s the ASEAN–4 countries experienced sharp increases in their external debts (Table 4.2), and much of the new debt was borrowed from private commercial sources. The external debt crisis that erupted in Mexico, Brazil, Argentina, and the Philippines has made international bankers and private investors wary of committing any more funds to the less developed countries. At the same time, budget constraints and domestic political problems in the OECD nations have made it difficult to increase capital infusions from official sources. With the slowdown in both private lending and official development assistance (ODA), coupled with mounting debt service payments, net capital inflows into the ASEAN–4 countries began stagnating in the mid–1980s. Total inflows of capital actually declined in South Asia between 1983 and 1985.

The slowdown in capital inflows is also a result of the unwillingness of the already heavily indebted ASEAN–4 countries to accumulate even more debt. The high cost of servicing commercial borrowing is one factor. Another is the inability of some governments, particularly the Philippines and Indonesia, but also Pakistan, Bangladesh, and Nepal, to raise enough revenue to meet local costs of development projects receiving ODA funds. This ''debt fatigue'' is unlikely to disappear in the next few years, and it complicates the tasks of moving away from austerity policies and stimulating these economies again. Lower interest rates may help somewhat, but none of these countries is eager to rely as much as before on foreign loans. Alternative sources of foreign capital that do not create additional debt obligations are therefore being given priority.

Conversion of Foreign Debt to Equity

Securing the dollar-denominated debt of developing countries or converting it to equity is a new trend. Worried financial institutions with large loans to developing countries are disposing of potential problem loans on secondary markets for prices as low as 70 cents on the dollar. Japanese companies have been particularly active in purchasing dollar debt in order to swap it for local currency needed to undertake direct investment in the developing countries. The strong yen has made such purchases of debt even more attractive for Japanese investors. Within the ASEAN–4, the debt-equity conversions have involved Japanese companies purchasing problem loans of the Philippines at sharp discounts from foreign (mainly U.S.) banks and then converting the debt into local currency at 80 to 90 percent of the local currency value. The foreign bank is able to clear its books of bad loans and recover most of the cash value, the investing company makes a foreign exchange savings of 10 to 20 percent, and the government concerned has its debt obligations trimmed and gains foreign exchange. This type of swap boosts foreign direct investment in the indebted countries, and may further enhance the balance of payments and help in promoting structural adjustment.

Governments are also attempting to boost foreign equity inflows directly by selling government-owned enterprises to private investors. The Philippines, in particular, is trying to attract foreign private investors to buy at steeply discounted prices a large number of enterprises, including hotels, farms, and industrial and manufacturing firms, that fell into government hands. Other countries are eager to privatize government enterprises in order to reduce budgetary deficits and attract foreign capital inflows.

The magnitude of inflows of this type is likely to expand in the near term, and the ASEAN–4 and South Asian countries will look to Japan and the NICs as the likely sources of such funds. In the following section we consider the role of direct foreign investment (DFI) in helping the Asian countries restructure their economies.

DIRECT FOREIGN INVESTMENT AND STRUCTURAL CHANGE

In the past fears of losing economic sovereignty made most Asian governments reluctant to open their economies completely to DFI. But in the present context, DFI appears to be a more attractive means of financing development than in the past because it avoids creation of new external debt obligations for the host government. Foreign investment is also attractive because it potentially transfers scarce technology and skills to the host economy while having a positive effect on the balance of payments.

Quantitatively, DFI has accounted for only a small share of capital formation in both East and Southeast Asian countries, but qualitatively it has played a major role in the development of industries and services in Hong Kong and, especially, in Singapore (Andersson and Ternstrom 1987). And although the amounts of DFI flows into Taiwan and Korea were quantitatively small (less than 10 percent of total investment), they had a disproportionate effect on trade and manufacturing exports (Ramstetter 1986). ASEAN–4 countries with outward-looking development strategies, such as Malaysia and Thailand, have been able to generate new exports from manufacturing industries through greater proportions of DFI. The Philippines and Indonesia have been less successful in using DFI for export diversification (Andersson and Ternstrom 1987). The main obstacles have been domestic economic distortions created by government interventions that excessively protect domestic markets and alter prices in a manner favoring capital-intensive techniques and activities.

South Asia has had negligible DFI, though in recent years India, Pakistan, and Sri Lanka have moved to increase the scope for DFI, particularly in export-oriented firms.

The NICs are very likely to receive a larger share of Japanese DFI than are the ASEAN–4 countries, India, or the other countries of South Asia or China. This is because the NICs have better infrastructure and more skilled and efficient labor, are politically more stable, and have stronger domestic economies. But because of its large size and its huge current account surpluses, Japan can still be expected to be the largest source of all forms of external capital, including DFI, for the Asian countries, especially the ASEAN nations.

The ASEAN–4 and South Asian nations may want to diversify their sources of external finance, particularly with regard to DFI, by responding to the NICs' new balance-of-payments position and ongoing structural changes. The NICs can be viewed as supplementary suppliers of external capital, expertise, and technology for renewed development in the ASEAN–4 and South Asian countries. While Japan is the major source of external finance, there is already substantial DFI in the ASEAN–4 from Hong Kong and Singapore investors. The desire of NICs like Taiwan and Korea to move upscale in industrialization creates a new and growing potential supply of DFI for the ASEAN–4 and South Asian countries.

Investment from the NICs could help transfer the skills and technologies of labor-intensive manufacturing that are needed if the ASEAN–4 are to become internationally competitive. A logical process of relocation of labor-intensive activities from the NICs and Japan to ASEAN–4 and South Asian countries, facilitated by DFI, would be desirable as a means of stimulating trade and growth. China's desire to attract DFI will add to the competition for this source of capital. But the difficulties of conducting business and the political uncertainties in China have limited DFI flows to rather low levels (James and Young 1987).

The following section examines how such a pattern of DFI and other private inflows from the new capital-surplus countries might affect regional trade patterns.

COMPARATIVE ADVANTAGE AND CAPITAL FLOWS

Over the last few decades, economic development in the Asia-Pacific has in general followed the "flying geese" pattern, in which a country's comparative advantages, and hence its export pattern, shift with its level of industrialization and development (Akamatsu 1962). As a country loses its comparative advantage in certain goods, it moves into production of other goods. The advantage of the leader country in the first type of goods shifts to another country, which follows the first country in producing and exporting those goods. Export patterns of these countries over time resemble the V-shaped, follow-the-leader pattern of flocks of geese, hence the name.

Comparative advantage and the structure of exports change as a country develops and accumulates physical and human capital (Balassa 1977). Most LDCs lack physical and human capital, so they concentrate on exports of resource-intensive or labor-intensive products in which, depending on their specific factor endowments, they are likely to have comparative advantage. As countries develop, they accumulate human and physical capital, and begin to industrialize and to export more capital-intensive and, eventually, technology-intensive goods.

Transfer of Industry

The flying geese pattern of development implies that production is transferred to those countries with the comparative advantage. In Asia, Japan is the richest and most technologically advanced nation, but it sorely lacks natural resources. Japan's comparative advantage thus lies in newer capital- and high technology-intensive products, such as supercomputers and other leading-edge technological products, and in services, such as financial services. Because of the accumulation of physical and human capital in the NICs, comparative advantages in certain skill- and capital-intensive goods have shifted in large measure from Japan to the NICs.

The NICs, also lacking natural resources, previously were large exporters of labor-intensive goods, such as textiles, garments, and footwear, and of human capital (skilled labor)-intensive goods. Now, with their highly skilled, motivated, and relatively low-cost labor forces providing a comparative advantage, more capital- and technology-intensive industries have been shifting production to the NICs, for instance, steel, shipbuilding, automobiles, computers, televisions, radios, and other consumer durables.

The ASEAN–4 countries, endowed with abundant natural resources, remain large exporters of resource-intensive products, such as petroleum, rubber, timber, food, and agricultural products. But as the NICs' advantage moves toward capital-intensive goods and away from labor-intensive goods, their comparative advantage in labor-intensive goods shifts to other countries with abundant labor. The ASEAN–4 and South Asian countries have abundant labor but lack capital and technology. As the NICs produce fewer labor-intensive goods, there is room for the South and Southeast Asian countries to increase production of such goods if they can obtain the needed capital and technological inputs.

Role of DFI

DFI plays a central role in this flying geese pattern of shifting comparative advantage. DFI, by transferring technology and expertise (human capital) as well as (physical) capital to areas that have abundant factor inputs and cost advantages, is a means of facilitating the flying geese pattern of development. It cannot fill the capital needs of a developing country by itself, but it is an important source of capital if it is used efficiently. Moreover, DFI does not add to the host country's debt burden and frequently helps to expand its exports. Not surprisingly, then, most developing countries are increasingly welcoming DFI.

Japan and the United States have been the major foreign investors in the Pacific region, but that situation may be changing. U.S. DFI may grow more slowly because of limited U.S. savings, large and persistent trade and budget deficits, and the weakening of the dollar. Furthermore, a large portion of U.S. DFI is in petroleum (largely in Indonesia), and the recent weakness in petroleum prices has discouraged investment (Naya and Ramstetter 1987). For at least the next several years, capital is likely to remain scarce in the United States, and U.S. DFI is not likely to expand quickly. The United States is in fact currently importing huge quantities of capital to finance its deficits. Improvement of its budgetary, trade, and savings performances would help ease U.S. demand for foreign capital and leave more resources available for the developing countries.

Complementarity and Competition

Because of the great diversity in resource endowments and levels of economic development in the Asia-Pacific region, there are a number of significant complementarities and competitions in the region.

Japan, as a high-income and technologically advanced nation, is a leading producer and exporter of newer capital- and high technology-intensive industrial goods, as well as of high-value-added services. In these areas Japan competes with the United States and advanced industrial nations in Western Europe, and not so much with other Asian countries. But there is fierce competition from the NICs in certain capital-intensive industries, and there may be increased competition from Hong Kong and Singapore in financial services as well. Japan, however, has few natural resources and must import oil, minerals, food, and agricultural products. Many of these resource-intensive imports come from the ASEAN–4, North America, Australia, and China.

The NICs are also relatively poor in natural resources, and must import oil, minerals, and agricultural products. Much of these imports come from the ASEAN–4 countries. Both Taiwan and Korea heavily protect their limited agricultural sectors in seeking to be self-sufficient in staple foods. They, along with Japan, could reduce agricultural protection and expand their imports from the ASEAN–4. China exports resource-intensive products directly to Hong Kong and Singapore, and indirectly to Taiwan and Korea. The NICs have advantages in the production of standardized finished products that they export to the ASEAN–4 and China, as well as to Japan, the United States, and the rest of the world. Hong Kong and Singapore have built up thriving service sectors with substantial foreign participation.

For the resource-rich ASEAN–4 countries, comparative advantage lies in resource-intensive primary products. The ASEAN–4 also produce intermediate goods, such as cement and timber, as well as labor-intensive goods. The ASEAN–4 countries, however, need more capital to accelerate their growth, which might be provided by Japan and the NICs.

The South Asian countries, though less rich in natural resources than the ASEAN countries, have even greater human resources. The strong labor cost advantages have been offset by cost-increasing policy interventions. Further economic liberalization will be necessary if the domestic sectors are to become internationally competitive.

China, at present, is the wild card of East Asia. Because of its size, it is potentially a very influential economic force. It could provide a market for consumer, intermediate, and capital goods exports. It also could compete for markets with resource- and labor-intensive exports, and for scarce capital in the form of DFI, commercial loans, and development assistance. At present, though, it appears that China will have far less effect on the region than one might surmise, given its size. China is reluctant to import in large quantities for fear of unsustainable trade deficits. Moreover, at present it does not pose a serious competitive threat in international trade because of its internal inefficiencies.

CONCLUSION

The balance-of-payments situation of countries in the Asia-Pacific region has changed drastically in the 1980s. Most developing Asian countries have improved their current account positions, but by different means. Most of the ASEAN–4 countries have been forced to cut expenditures and growth to stabilize their external payments. Their export earnings have fallen with the decline in commodity prices, so that imports have had to be reduced even more. The ASEAN–4 are in a difficult situation with regard to external debt. The NICs and Japan, in contrast, have moved into surplus positions in their current accounts and, with the exception of Korea, have little external debt. Their exports have not fallen as much as those of the resource-rich ASEAN–4 countries.

There may be room for expansion of regional trade and investment flows that could serve to stimulate renewed growth in Southeast Asia and help to sustain growth of South Asia, the NICs, and Japan. DFI flows from Japan and the NICs to Southeast Asia and China could give a boost to these economies and to their exports without adding to their external debts. This could also facilitate the relocation of labor-intensive manufacturing activities in line with the comparative advantages of the various countries.

The success of the process of renewed economic development along the lines suggested here depends on the maintenance and expansion of international trade and investment. Restrictive trade practices could slow down or even reverse the process.

FIVE

Property Rights, and Petroleum Exploration and Development in the Asian Developing Countries

WIROTE MANOPIMOKE AND
WILLIAM E. JAMES

INTRODUCTION

The energy crisis created by the Arab oil embargo in 1973–1974 resulted in a considerable change in investment behavior among multinational oil companies. According to a U.S. Department of Energy report on the investment behavior of 26 major U.S. energy companies,[1] it was found that these companies had reduced their dependence on Middle East oil by shifting to other sources of oil and gas reserves. After 1974, 80 percent of these 26 companies' capital spending was allocated to petroleum operations. While their refinery expansion and investment in the direct marketing of finished petroleum products slowed, spending on upstream operations of exploration and development increased dramatically.

A study of the industry's petroleum exploration effort also was undertaken by the M.I.T. Energy Laboratory. The study shows that 78 percent of calculated expenditure, using Chase Manhattan Bank data allocated by rig time, was in developed countries, 16 percent in oil-exporting developing countries (OXDCs), and only 6 percent in oil-importing developing countries (OIDCs). Investment in the United States itself accounted for 54 percent of total outlays.

The relative success of investment in oil and gas development can be measured in two ways: (1) according to the quantity of oil and gas discovered per wildcat drilled or (2) according to the cost per barrel added to reserves. Table 5.1 shows these measures of exploration efficiency. Column 1 is the ratio of gross oil reserve additions to the total number of exploratory wells, and column 2 is the ratio of gross oil and gas reserve additions to the total number of exploratory wells. On the average for the world, 2.7 million barrels of oil equivalent were discovered per exploratory well. In the OXDCs this rate was 42.5 million barrels of oil equivalent, and in the OIDCs it was 4.3 million barrels. Only 0.6 million barrels of oil equivalent were discovered per wildcat in the developed countries.

Table 5.1
Exploration Effectiveness, 1975–82

	Million barrels per well	Million barrels of oil equivalent per well	1980 US$/ barrels of oil equivalent
United States	0.21	0.47	5.72
Canada	0.16	0.61	2.75
N. Sea countries	3.33	10.98	2.31
Other developed countries	0.45	1.09	14.54
Developed countries	0.24	0.61	4.95
Middle East OXDCs	143.56	189.50	0.11
Asian OXDCs	5.59	12.34	0.68
Asian OIDCs	3.70	8.06	1.57
Asia	32.42	46.28	0.26
African OXDCs	14.21	16.70	0.84
African OIDCs	1.47	3.73	1.19
Africa	11.29	13.73	0.86
Latin American OXDCs	28.26	35.92	0.31
Latin American OIDCs	1.17	2.90	1.44
Latin America	13.58	18.01	0.41
World Total	1.78	2.70	1.31
Developed countries	0.24	0.61	4.94
OIDCs	1.82	4.25	1.48
OXDCs	31.76	42.53	0.29

OIDCs = Oil-importing developing countries.
OXDCs = Oil-exporting developing countries.
Source: Blitzer et al., 1985, p. 296.

In terms of total investment expenditures to gross oil and gas reserves added, the level of investment is higher in developed countries despite the higher costs of finding new reserves. On average, it took $1.31 to discover a barrel of oil equivalent worldwide, compared with only $0.29 in the OXDCs, $1.48 in the OIDCs, and $5.00 in the developed countries (Blitzer et al. 1985).

The evidence seems to indicate that there is a gap between the developing and the developed countries. In Asia, perceptions of such a gap may have been lessened by the boom in petroleum exploration in Southeast Asia after the first oil shock. The amount of foreign investment by oil companies in Malaysia, Indonesia, Brunei, Thailand, and the Philippines rose substantially. Yet even in Indonesia, where investment has been greatest, only about one-third of tertiary basins have been explored, and estimates of reserves remain very conservative (Economist Intelligence Unit 1986, p. 66).

It now seems unlikely that exploration activities will resume at anywhere near the pace of the time period from the second oil shock to 1985. Downward revisions in price expectations caused by the weakening of OPEC, the rise in non-OPEC production capacity, energy conservation and substitution, slower economic growth, and the general perception of weakness in prices of almost all commodities will dampen investment in exploration considerably. Still, interesting questions can be posed: Why has the search for oil and gas been so concentrated in the more developed countries, despite the prevailing higher costs? Could the relatively low level of investment in petroleum exploration in the developing countries be explained by differences in geological risk, political risk, and tax-contractual arrangements?

Geological risk can be ruled out as a complete explanation for low investment in petroleum exploration in developing countries. The data in Table 5.1 show that the per barrel cost of adding new reserves is substantially lower in developing countries than in developed countries.

The gap in exploration between developed and developing countries was argued by some to be a case of failure of private market forces. Under that reasoning, it was argued, in the wake of the second oil shock, that a special energy facility should be set up under World Bank auspices. (The proposal was scrapped when the United States let it be known that it would oppose providing additional funds to such a facility.)

A fuller explanation of the gap would need to incorporate elements of political risk and differences in tax-contractual arrangements, and to assess more broadly the policy and institutional factors that might hold back investment commitments needed to effectively explore the energy potential of the developing countries.

BEHAVIOR OF FIRMS FACING RESTRICTED RIGHTS

We hypothesize that institutional arrangements are prominent in determining the different levels of oil and gas exploration and development among countries. The institutional approach used here argues that multinational oil companies and governments interact symbiotically. While nation-states may own the natural resources on land and under their offshore waters, they will normally lack the means to explore and develop those resources. They must harness the energies and talents of private companies to that end.

Countries like the United States, Canada, and Australia grant rights directly to private companies by fixing legislation in advance. The right to explore and/or exploit petroleum resources may be granted under standard licenses and leases. Royalty taxes and other payments to be made are also determined by legislation. However, in developing countries the system of legislation in effect leaves it to the government to grant rights to explore and/or to exploit petroleum resources on the basis of individually negotiated agreements.

A major theme in the institutional approach is how changes in government policy and/or legal entitlements to resources modify private investment decisions.

The law may be viewed as a means for the state to alter, transfer, or restrict the conditions and rights of ownership to a set of goods. The legal system and policy environment have important implications for efficiency and the distribution of welfare. Coase (1960, p. 16) points out that once property rights are defined in a particular manner, transaction costs may preclude altering the established rights in order to improve efficiency:

> . . . One arrangement of rights may bring about a greater value of production than any other. But unless this is the arrangement of rights established by the legal system, the costs of reaching the same result by altering and combining rights through the market may be so great that this optimal arrangement of rights and the greater value of production which it would bring, may never be achieved.

Uncertainty regarding future changes in property rights can also have strong effects on incentives to invest. Alteration of property rights can change the value of an asset and the terms of trade by placing restrictions on its use and exchange.

Since the search for oil and gas has been so concentrated in the developed countries, where expenditures to add a barrel of oil equivalent to reserves are relatively high, technical and geological conditions seem not to be the major factors in the oil companies' investment decisions. In effect, differences in political, legal, and institutional environments, and specific economic policies and conditions (such as infrastructure, availability of amenities, and, in some cases, subsidies), are the major elements explaining the exploration and development gap between developed and developing countries.

The benefits that a petroleum company might expect from investment in exploration equal the present value of the rights to a stream of future earnings. The certainty with which the companies anticipate this stream of return, the prevailing interest rate, preferences for risk, and other factors (price expectations) will enter into the valuation of the investment. The legal framework is relevant because it affects financial return. Profitability and the company's other interests—such as control over operations, freedom to make production-level decisions, pricing, and marketing—are all important factors determining the expected return. Perceptions of delays or difficulties in clearing imports of equipment and materials by lengthening the payback period would also reduce investment commitments.

In addition to the costs of physically searching for oil and gas deposits are the transaction costs of doing business with a foreign government. Transaction costs are defined here as the costs of specifying and enforcing the contracts that underlie exchange and transfer, and comprise all the costs of political and economic organization that permit agents to capture gains from trade. Transaction costs can be divided into ex ante and ex post costs. Ex ante costs are those of drafting, negotiating, and safeguarding agreements. Ex post costs are (1) maladaption costs incurred when transactions drift out of alignment; (2) the haggling costs incurred if bilateral efforts are made to correct ex post misalignment; (3) the setup and running costs associated with the governance structures to which disputes are referred; and (4) the bonding costs of effecting secure commitments (Williamson 1985).

Differences in the legal system and institutional environment incorporating all the above economic and political factors can result in varying benefit-cost ratios, and hence will affect the company's choice of investments. Assessment of the viability of investment projects for oil companies involves three main factors: geological, fiscal, and institutional. Most of the literature has been concerned with either geological or fiscal factors, or both. In this chapter we examine the institutional factor as well.

EVIDENCE

The discussion that follows will concentrate on petroleum exploration activities and investments among Asian developing countries (ADCs). Petroleum exploration and development activities in the region are undertaken by foreign multinational oil companies (usually privately owned) and domestic state-owned oil and gas companies. The contractual arrangements for petroleum in the region represent a spectrum of global contractual conditions. They include concession agreements operating in Brunei, Pakistan, Taiwan, Thailand, and South Korea. Production-sharing contracts predominate in Indonesia, India, Malaysia, Burma, Bangladesh, and Sri Lanka. Risk service contracts can be found in the Philippines and Vietnam. Joint-venture participation arrangements exist in Indonesia and China.[2]

Tables 5.2–5.4 present the exploration activities in the region. The country rig profile (Table 5.3) shows that rig activity was largely concentrated in five countries—Indonesia, India, Burma, Brunei, and Malaysia—that have high proven oil or gas reserves. However, it should also be noted that India and Burma mostly conduct their own exploration through state oil companies. Thus, the high level of activity in these two countries did not result from foreign investment.

Even though Bangladesh and Pakistan have large proven reserves of natural gas, they have not been successful in attracting foreign oil companies. Drilled wells in Bangladesh fell by more than 50 percent during 1976–1979. (See Table 5.4.) Pakistan showed a slow increase in activities even though the proven reserves were not much different from those of Malaysia. The Philippines experienced a downturn in all activities in the period 1976–1984, while Thailand experienced an upturn during the same period.

Political Risk

The case of the Philippines may reflect the effect of political risk on upstream investment. The decline of exploration there in 1983 was surely related to political uncertainty. An economic and political crisis engulfed the country following the assassination of former Senator Benigno Aquino in August 1983. Oil and gas exploration in the Philippines remained stagnant until 1986. The prolonged eco-

Table 5.2
Geophysical Activity in Developing Asia, 1973–84 (seismic party months)

	Bangladesh	Brunei	Burma	India	Indonesia	Malaysia	Pakistan	Philippines	Thailand
1973	na	5	18	295	324	7	18	6	5
1974	0	na	na	300	308	na	39	na	5
1975	8	na	na	305	234	0	82	na	3
1976	16	14	na	312	145	5	48	15	1
1977	26	8	na	312	83	22	48	23	2
1978	19	0	na	288	140	2	37	38	0
1979	20	6	na	294	210	2	82	53	14
1980	40	11	18	300	250	16	87	45	55
1981	24	9	18	294	270	5	89	26	31
1982	23	3	19	355	299	3	75	16	65
1983	23	18	15	403	184	3	76	27	42
1984	31	6	12	414	176	12	61	20	23

na = Not available.

Source: Energy Economics Research, Ltd.

Table 5.3
Estimated Number of Active Rigs in Developing Asia, 1977–86 (end of period)

	Bangladesh	Brunei	Burma	India	Indonesia	Malaysia	Pakistan	Philippines	Thailand
1977	2	7	17	43	46	8	5	6	1
1978	3	8	18	40	62	11	8	6	2
1979	2	8	22	40	66	13	13	15	2
1980	3	8	26	43	77	12	14	12	5
1981	5	7	30	44	87	12	15	13	7
1982									
Land	5	3	32	41	60	0	17	12	3
Offshore	0	7	1	10	32	13	0	3	5
1983									
Land	2	3	35	41	62	0	17	7	6
Offshore	0	7	1	14	25	10	0	1	6
1984									
Land	4	2	32	40	56	0	17	3	5
Offshore	0	7	1	17	26	9	0	0	6
1985									
Land	5	1	32	49	54	0	17	1	3
Offshore	0	5	1	15	20	9	0	0	3
1986									
Land	5	1	32	74	25	0	18	0	2
Offshore	0	3	1	21	15	7	1	0	1

Source: Energy Economics Research, Ltd.

Table 5.4
Wells Drilled in Developing Asia, 1970–85

	Bangladesh	Brunei	Burma	India	Indonesia	Malaysia	Pakistan	Philippines	Thailand
1970	na	40	3	na	84	18	7	2	na
1973	na	40	4	5	501	17	2	11	8
1975	na	50	10	na	588	11	5	13	10
1976	5	50	9	na	513	51	5	8	17
1977	5	75	15	30	409	57	11	15	3
1978	2	77	40	119	536	89	13	18	8
1979	2	50	68	125	636	158	16	29	10
1980	4	56	60	45	550	107	18	25	25
1981	5	38	60	126	727	99	24	15	26
1982	5	18	76	153	773	91	22	15	45
1983	1	69	150	170	891	96	34	4	96
1984	2	63	237	193	762	103	42	3	108
1985	2	38	209	217	678	61	76	na	92

na = Not available.

Source: Energy Economics Research, Ltd.

nomic crisis was not ameliorated by the revolution of 1986. However, it is not easy to distinguish the point at which the seriousness of the political situation will hinder exploration activity. In 1978, following the end of the Vietnam War and the unification of that country, three foreign oil companies—Agip (Italy), Bow Valley (Canada), and Deminex (West Germany)—signed contracts to conduct offshore petroleum explorations, despite the great political uncertainty. Assuming that the firms are oriented to maximizing profits, we might argue that if the expected net returns from investments are high enough—even with high political risks—the firms will invest. It is the comparative stream of future income (after taking into account the political uncertainties) that the oil companies use to decide whether to invest in any country.

Contractual Form and Tax Treatment

Generally, past economic research in oil and gas exploration and development explains the low foreign investment in LDCs as a problem of nonoptimal tax-contractual arrangements. (See, for example, Blitzer et al. 1983.) However, while almost all of the exploration and development contracts are long-term agreements, long-term contracts are rarely considered in standard microeconomic analysis. As Alchain and Woodward (1987, p. 1) describe it:

> The traditional price-theoretical paradigm in which gains from trade depend on comparative advantage in production and buyers and sellers who have equal information is distinguished by two absent features. First, no one is relied upon for directions about what to do; market prices direct production and exchange. Second, long-term contracts, which contain some future options of the contractual parties, have no role. Yet, every modern society violates both conditions.

The assumption of neoclassical economics is that all costs are adjusted as conditions change. Labor and capital are reallocated to their highest-value uses. In addition, neoclassical economics often assumes that transaction costs are negligible and that adjustments are immediate, or nearly so. The limitation of the received theory regarding the issue of long-term contracts is indicated by the fact that most of the existing studies only alter an excise tax in order to obtain the highest present value from different contractual forms. The logic is that if competition to encourage foreign investment exists among the host countries, then royalties or the tax rate should not vary much across countries.

The prediction is that competition will drive developing countries seeking foreign investment in exploration to adjust the contract terms so that the expected return is similar. The neoclassical model does not say anything about how long the adjustments will take or about the political and other constraints that governments may be operating under, thus limiting its explanatory power. The model can perhaps be usefully extended to incorporate the transaction costs associated with negotiation and implementation of long-term contracts.[3]

The institutional approach attempts to explain differential success in attracting foreign investment according to differences in transaction costs. Transaction costs involved in long-term contracts can be divided into ex ante and ex post costs. Ex ante costs are the costs of drafting, negotiating, and safeguarding agreements. Contracts can be drawn up with a great deal of care as complex documents in which numerous contingencies are recognized, and appropriate adaptations by the parties are stipulated and agreed to in advance. Or the contract can be very incomplete, with the gaps to be filled in by the parties as contingencies arise. Therefore, rather than contemplate all conceivable bridge crossings in advance, actual bridge-crossing choices are addressed only as events unfold. The latter course would involve lower ex ante but possibly higher ex post costs.

The ex post costs can be described as follows: Suppose that the contract stipulates x, but, with the benefit of hindsight (or in the fullness of knowledge), the parties discern that they should have done y. Getting from x to y may not be easy. The manner in which the associated benefits are divided is apt to give rise to intensive, self-interested bargaining, and complex strategic behavior may be elicited. Referring the dispute to another forum may help, but that will vary with the circumstances. An incomplete adaptation will be realized if, as a consequence of efforts of both kinds, the parties move not to y but to y' (Williamson 1985).

A complicating factor in all of this is that the ex ante and ex post costs of contracting are interdependent. Put differently, they must be addressed simultaneously rather than sequentially. Also, costs of both types are often difficult to quantify. The difficulty, however, can be mitigated by assessing transaction costs in a comparative institutional way, in which one mode of contracting is compared with another. Accordingly, India is shown as an example of ex ante costs; Burma, Indonesia, Malaysia, and Thailand are shown as examples of ex post costs.

Petroleum activity in India is dominated by India's state oil companies, the Oil and Natural Gas Corporation and Oil India, Ltd. Prior to 1971 only one foreign contractor, Burmah Oil of the United Kingdom (in a joint venture with the government), was active in drilling. Until the mid–1970s, the Indian government sought to rely entirely on state oil enterprises. However, during 1974–1975, the government signed production-sharing contracts with three foreign companies, all of which withdrew after a limited amount of offshore drilling.

In September 1978, India decided to revive the existing policy of exploration and issued an invitation to companies to show "interest" in the exploration and development of hydrocarbon resources. Despite the lack of details about the specific areas being offered and the terms of the contracts, the advertisements drew responses from 67 companies by the deadline date.

The government of India released a "short list" of 34 companies in December. They were invited to take part in discussions of a technical nature. Information released with the "short list" on the terms of the contracts confirmed that they would be for production sharing. The contractors would be required to sell "cost oil" to the government at a fair price, and preference in awarding contracts would be given to those companies offering to sell their entire share of "profit oil" to the government.

Only 22 companies stayed on for further discussion with the Petroleum Ministry. The negotiations dragged, in part because the ministry had not prepared clear-cut terms and conditions for the tender.

Another hurdle in coming to an agreement with foreign oil companies was a guideline issued by the Petroleum Ministry that required the companies to surrender their concession five years after the discovery and commencement of the production of oil or gas. This was unacceptable to most companies; they argued that they could not recoup their investment in such a short time (*Petroleum News* 11, no. 10, 1981).

In August 1980, India's government again invited foreign oil companies to explore for oil and gas on the basis of a production-sharing formula. Protracted discussions were carried on with a number of international companies, but contracts could not be negotiated under the terms offered by the government, which were similar to those provided in the three production-sharing contracts signed earlier. Finally, the government relented on the price issue and, although foreign companies would not be allowed to export oil until India became self-sufficient, the petroleum minister announced that India would buy the companies' oil shares at international prices. It is also reported that India offered to extend the terms of the contracts. In December 1981, a contract was signed with Chevron Overseas Petroleum for an offshore block. India has had little success in attracting exploration by foreign petroleum companies, and Chevron was the only non-Indian company conducting exploration early in 1983. India's contracts are reported to be unattractive to the industry. In 1985 Chevron withdrew from further exploration activity after investing a total of $29 million (EIU 1986). Many of the blocks being offered for contracts are in very deep water where drilling costs are high.

Ex post costs have also arisen in many ADCs. More specifically, the ex post costs occur after the host countries and oil companies have reached an agreement for petroleum exploration and development, and the oil companies are in operation. The ex post problem, therefore, is one of contract enforcement. Even though a contract may be enforceable, it is still subject to postcontractual opportunistic behavior. The most extensive treatment of postcontractual problems emphasizes two universal behavior traits. First is "bounded rationality," which means that people are limited in the information they have and in their ability to process it. This implies incomplete information about market opportunities, limited ability to predict the future and to derive implications from predictions, and limited ability to prespecify responses to future events. Second, people are opportunistic. This trait combines bounded rationality with self-interest. When there is conflict between what is in their interest and what they have agreed to do for others, they will act more in their interest and make the detection of their action and enforcement of promised performance more costly (Alchain and Woodward 1987).

The possibility of opportunistic behavior is real. Once all assets are in place (the wells drilled), the oil companies' properties are locked into the host country. These specific producing assets are, therefore, hostage to the host country. Ex post problems often arise from trying to appropriate economic rent from natural resource discoveries. They often lead to a slowdown in exploration and development.

In Burma, the last four foreign offshore exploration groups pulled out in 1977. Since the oil industry was nationalized by the government, oil exploration has been a monopoly of the state that is operated by Myanma Oil Corporation. Thus, all foreign companies' exploration activities were under production-sharing contracts with Myanma. An interesting commentary was made in an official report on expenses incurred by service contractors. While Myanma spent an average of 1,076 kyats ($165) per foot for the 12 offshore test wells drilled, the contractors spent 4,058 kyats ($624) per foot for 17 test wells. The contractors' greater costs were to some extent due to their employees' higher pay and travel expenses between the Singapore base and Rangoon. However, their expenditures also included certain items that could have been avoided, such as demurrage charges borne by the contractors for the delay in clearing a consignment of barites from the port of Rangoon (*Petroleum News* 11, no. 10, 1981). Burma has more recently received assistance from the Japanese-owned Burma Petroleum Company in ongoing offshore exploration.

The violation of petroleum contracts by the Indonesian government is another example of the effect of ex post costs on the exploration activities of foreign oil companies. Following the sharp rise in oil prices in 1973, the Indonesian government demanded that the contracts be renegotiated to increase Pertamina's share of the output. The action was resented by the companies, although it apparently did not reduce their profits to the point where they were unwilling to negotiate additional contracts.

In March 1976, as a consequence of Pertamina's chaotic financial affairs, the Indonesian government violated its contracts with the foreign petroleum companies. In April 1976, the government reduced the estimated profit margin of the largest producer, Caltex, by $1 per barrel (about $300 million per year) by renegotiating its work contract. In the case of companies operating under production-sharing contracts, the government raised its 65 percent share of oil left after cost recovery to 85 percent and increased the period over which companies could recover their costs from 7 to 14 years. On top of these developments, in April 1976 the U.S. Internal Revenue Service ruled that the output shares of foreign governments under production-sharing contracts were to be treated as royalty payments and not allowed as credits against U.S. tax obligations. As a consequence, petroleum companies sharply reduced exploration commitments in Indonesia in 1976 and 1977 (Mikesell 1984). Table 5.2 shows that the geophysical activity in that country fell dramatically in 1976 and fell even further in 1977. Also, wells drilled showed a declining trend during the same period (Table 5.4).

In Malaysia, a slowdown of foreign investment occurred in 1976. The Malaysian petroleum legislation of 1975, particularly the clause that would give Petronas, the national oil company, the right to buy a new class of ''management shares'' carrying majority voting rights, met with concerted opposition from the foreign oil companies. Exxon refused to commit any further expenditure and stopped further exploration, and Conoco withdrew its drilling rigs operating in Malaysian waters (*Petroleum Economist* 43, 1976). Later Petronas purchased Conoco's rights and investments, and formed its own exploration subsidiary. Since then, Malaysia has been quite successful in negotiating exploration agreements with Esso, Shell, British Petroleum, and other companies. The adoption of a consistent set of policies since the mid–1970s is cited as a reason (EIU 1986, p. 111). It also is directly participating in exploration activities. Malaysia has been able to increase its estimated reserves of oil, and Petronas expects to grant two to three new concessions per year during the rest of the 1980s (EIU 1986, pp. 110–111).

The last case is Thailand. Petroleum exploration in Thailand has gone smoothly since natural gas was discovered in the Gulf of Thailand. However, in 1984, Thai exploration activity was hindered for the first time by disputes between the government and the foreign oil companies over such basic questions as pricing and production. The output of the Erawan field of Union Oil Company became a controversial issue because of the great discrepancy between its estimated resources and its actual outputs. However, the real problem with an exploration slowdown that Thailand might face in the near future is the lengthy negotiation over gas pricing that the foreign oil companies might anticipate as a high ex post cost. There also has been some onshore exploration for petroleum in northern Thailand by foreign oil companies, which resulted in only one commercially viable deposit.

CONCLUSION

This study has attempted to show that when foreign oil companies evaluate their exploration investments in developing countries, transaction costs can be a significant factor. These costs are created by the legal system and institutional setup, and are difficult to quantify. From the evidence it is not possible to conclude that such costs are especially prominent in attracting or discouraging foreign investment in oil and gas exploration. Expectations about future oil prices that have more impact on profitability are probably the dominant factor, though we have not directly examined this dimension.

Ecological considerations, political risks, and nonoptimal tax-contractual arrangements can also be used in part to explain the differences in foreign investment in exploration and development among different countries. Using the institutional approach, it appears that a well-defined and enforceable legal system can be useful in attracting foreign investment, but is hardly necessary or sufficient. Nevertheless, the better defined and enforceable the law is, the lower the cost of doing business.[4]

Political stability, well-defined legal and institutional arrangements, good geological potential, and competitive contract terms can help to attract foreign investment in exploration activities. Yet, with the less bullish oil price expectations of recent years, OIDCs still may not attract the levels of investment they would like from private companies. With the long time lag between exploration and subsequent commercial development, Asian OIDCs may once again find themselves in a bind should oil prices unexpectedly rise sharply. The ASEAN oil exporters could be in a less advantageous position than otherwise because of more limited capacity to export. Still, neither group of countries is in a position to gamble much of its own funds on further exploration activities. The best that can be hoped for is greater stability in prices and expectations thereof while suitable improvements are made in institutional conditions.

NOTES

1. The following is the list of companies: Amerada Hess; American Petrofina; Ashland; Atlantic Richfield; Burlington Northern, Inc.; Cities Service; Coastal; Conoco; Exxon; Getty Oil; Kerr-McGee; Marathon; Mobil; Occidental; Phillips Petroleum; Shell Oil; Standard Oil Company (Indiana); Standard Oil Company (Ohio); Sun Company; Superior; Tenneco; Texaco; Union Oil of California; and Union Pacific. The merger of Cities Service with Occidental Petroleum in 1982 reduced the number of reporting companies from 26 to 25 beginning in 1983, and the 1984 mergers of Chevron with Gulf Oil, Mobil with Superior Oil, and Texaco with Getty Oil reduced the number of reporting companies to 22 beginning with the 1985 reporting year.

2. The concession basically gives all production to the concessionaire while imposing commensurately high tax and royalty rates. The production-sharing contract divides production between the government and the contractor after allowing a portion for cost recovery. It imposes a lower income tax than in concession contracts and sometimes includes a royalty.

These different types of contracts represent the intention to have different risk-sharing arrangements between the host country and the foreign oil companies. However, there are complex variations in each contract, and the implications of property rights embodied in the contract form have to be taken into account.

3. For discussion of these issues, see James Roumasset, "Agency Theory, Agricultural Firms, and the New Institutional Economics," Department of Economics, Working Paper no. 1, University of Hawaii, Honolulu, August 1988.

4. Transaction costs alone cannot explain all of the gap in investment between developed and developing countries. Wallis and North (1987) point out that profit-maximizing firms willingly incur transaction costs as long as marginal benefits of transactions exceed marginal costs. It is not the existence of transaction costs that deters investment; it is the risk of unexpectedly incurring substantial losses in dealing with a sovereign state. Even if transaction costs are high, investment will be forthcoming as long as they are stable and predictable (and expected returns are high enough).

PART II
Country Case Studies: The Pacific

SIX

The Energy Sector and Economic Issues in New Zealand

GOVIND P. SAHA

INTRODUCTION

The energy sector in New Zealand, with the exception of the oil market, is dominated by agencies in the public sector. In addition, the government has, in the past, adopted a hands-on approach in the management of the energy sector. This process was strengthened following the two oil crises in the 1970s, which led to a belief that energy security and economic efficiency would be enhanced by government's assuming the role of coordinator in the energy markets, and that this would be in the national interest. The government placed heavy emphasis on self-sufficiency in all fuels, especially the liquid fuels that were then largely imported. An integrated energy planning process at the national level was initiated in 1980. Its purpose was to ensure that energy supply and demand were properly and adequately matched, and that any adverse impact on New Zealand's economy resulting from a shortfall in the supply of energy would be avoided. It was also believed that, given the long lead times in energy projects, the coordinated energy planning process would result in the most efficient allocation of resources. In this mode of operation, the government not only set the environment in the energy markets but also was the largest trader/investor in the energy market.

This chapter describes the fundamental changes now taking place in New Zealand's energy sector. It briefly considers some of the recent structural and regulatory changes in the various markets of the energy sector and then examines in some detail the issues in the electricity markets.

RECENT STRUCTURAL AND REGULATORY CHANGES

Since July 1984, when the Labor government took office, there has been a major shift in the government's approach to the management of the energy sector.

The energy planning approach has been replaced by a greater reliance on the market as the best means of driving decision making. This in turn has dramatically changed the emphasis of the government's role in the management of the energy sector.

Instead of looking at specific development or operational decisions, the thrust of government work is now to create an appropriate environment for others to make decisions. For the energy sector this has meant two major policy thrusts, which are encapsulated in the key words "restructuring" and "deregulation." Although the pattern of activity in New Zealand is similar to that of other Western countries, the extent and speed with which the economic and energy reforms have been implemented are quite remarkable and unprecedented.

Restructuring of the energy sector is occurring at three levels. The primary driving force recently has been the government's program of corporatization (the process of separating commercially driven trading activities within the government's control and transforming them into state-owned companies based on the private sector model). In the energy sector, the process has seen the termination of the state coal mines and electricity divison of the Ministry of Energy, and their replacement by the new entities Coalcorp and Electricorp. These new corporations have joined the existing state-owned enterprise (SOE), Petrocorp, although Petrocorp is now changing its status as a result of the decision to sell 30 percent of its shares to the public. For many of these SOEs, it is clear that corporatization is only a step toward partial or full privatization.

The change in status of these trading activities is shown in Figure 6.1. It is especially noticeable that these three SOEs have a dominating influence over most of the energy sector. The only exception is the oil market, which continues to be dominated by overseas oil companies—Mobil, Shell, BP, and Caltex.

Although the new corporations are based on the private sector model and thus are limited-liability companies under the Companies Act, they are subject to special accountability and control provisions under the State-owned Enterprises Act. Chief among these provisions are the requirement to agree to a statement of corporate intent with the shareholding ministers that is published and can be debated in Parliament, and the provision allowing the minister of finance to set dividend rates.

One of the key questions being addressed in regard to these SOEs is how and by whom their activities should be monitored. A discussion paper on this subject has been released by the government, and public submissions are being sought. The discussion paper is mainly focused on the objectives and content of commercial monitoring, which is primarily designed to be a surrogate for the disciplines of the share market (a means of ensuring that the shareholders' investment is being properly rewarded).

In addition to the structural changes, major shifts in the regulatory environments are taking place for all the energy markets. For example, from January 1, 1988, there will be virtually no regulation or control in the electricity generation market. In other words, there will be no restrictions to entry and no price

Figure 6.1
Corporatization in the Energy Sector

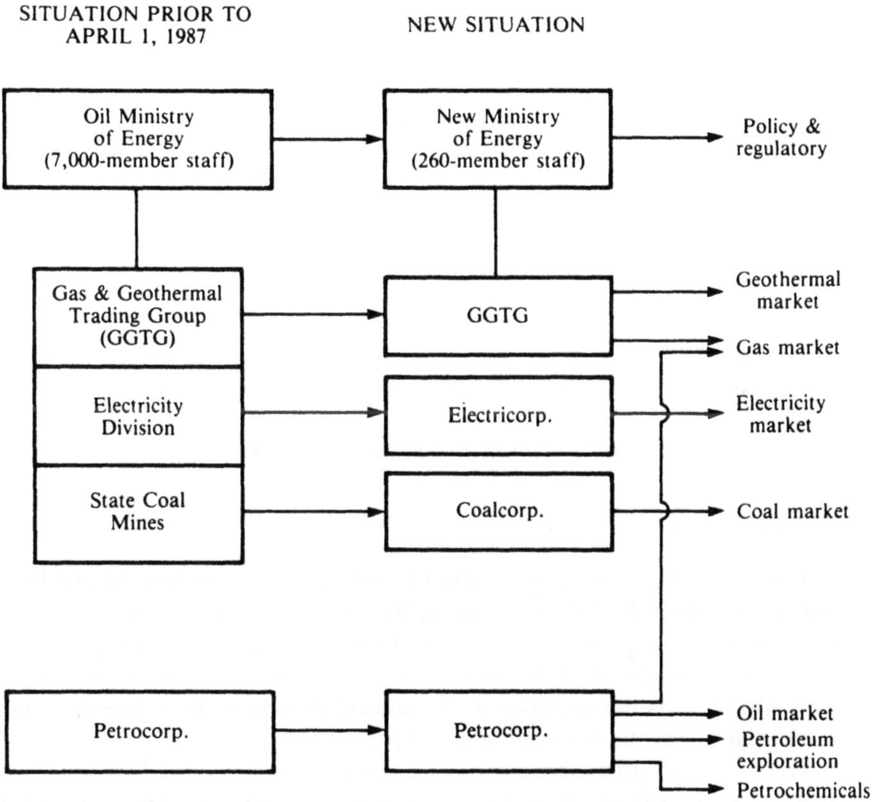

Source: Govind P. Saha, Ministry of Energy, Wellington, New Zealand, 1987.

controls on the electricity bulk supply. Preventing or lessening competition or new entry would be illegal under the new Commerce Act, and therefore subject to legal sanctions. The government intends to introduce similar deregulatory measures in the oil and gas sector, and details are being finalized. The coal sector is already substantially deregulated.

RESTRUCTURING ENERGY SUBSECTORS

The Crown's interest in the energy business is extensive, and the convolutions of its interactions have contributed to sector inefficiencies. Therefore, a major effort is being made to restructure interests to reduce the level of direct government involvement and to create a more open and competitive market. The problems and objectives in each area are briefly canvassed below.

Geothermal Market

The geothermal market is characterized by almost total domination by the Crown, through either the ministry's Gas and Geothermal Trading Group (GGTG) or Electricorp. There are no large-scale private sector participants, although a number are potentially interested. From a restructuring perspective, the key question is whether the government should continue to be directly involved in geothermal through GGTG. The options are to retain the existing system, transfer the Crown's activities to an SOE, or sell the business to the private sector. At present, no decision has been made.

Coal Industry

The coal market is the one energy subsector where no explicit restructuring is required. It is extremely competitive, both internally and with other fuels, so there is no reason not to allow Coalcorp to seek its own commercial destiny. The Crown has no other direct involvement in the coal industry.

Oil Industry

The oil industry represents a distinctly more complex restructuring problem. The key components in the industry are the synthetic gasoline plant (Synfuels), the Marsden Point refinery, and imported product. The Crown is implicitly or explicitly involved in all three areas. It is a part owner of the Synfuels plant, it basically supports the operation of the refinery through various measures, and it effectively controls the importation of product.

From a restructuring point of view, the key to the system is the Marsden Point refinery. (The synthetic gasoline plant is more of a problem from a gas sector point of view, and is thus discussed further in the next section.) Discussions currently under way with the New Zealand Refining Company are aimed at providing incentives for the efficient operation of the refinery from a systems point of view, including the opening up of the market to competition from product imports. A solution that allows product imports is most desirable, as it would effectively open up the New Zealand market to the rest of the world. Solutions that protect the refinery from import competition, but provide incentives for improved internal efficiency, are less desirable but may prove to be practicable.

Achievement of refinery restructuring is considered to be the key to successful deregulation of the oil market, since the protected position of the refinery gives the existing oil wholesalers a potential competitive advantage.

Natural Gas Industry

If the oil industry represents a difficult restructuring problem, the natural gas sector is more difficult still. The key state-owned institutions in the sector are

Petrocorp, the GGTG, and, through the GGTG, the Synfuels venture. It is interesting to note that the Synfuels operation at Motunui has a rather passive role in systems operations. It is purely a processing operation run on a tolling basis, and is restricted to arrangements at the wholesale and bulk use level. Downstream of the main transmission network are the major utilities that distribute natural gas at the local level.

The key problem in restructuring the gas sector is not so much the institutions themselves as the contracts that govern their operations. A series of complex contractual relationships that govern the development and utilization of the Maori resource have been established. These contracts are a legal nightmare and have established long-term relationships that, in many instances, have turned out to be less desirable than first thought. The contracts fall into three broad categories:

- Those concerned with the Maori venture, including the arrangements among the Maori partners, especially the gas sales contract with the Crown that gives the Crown monopoly control over gas supply
- Those concerned with the Synfuels venture, including the venture arrangements between Mobil and the Crown for the Motunui plant, the arrangements with Mobil for use of the technology, and the role of the Crown in product marketing
- Those concerned with sale of gas to bulk users, especially Petrocorp (including NGC, Petralgas, and Petrochem) and Electricorp.

The complexity of sorting out these contracts is such that the problem has thus far been put in the "too hard" basket. However, this area is of priority concern to the Ministry of Energy because the ministry's one remaining trading activity—the GGTG—is a key participant in the sector. The ministry's view is that the way into the tangle is through the Synfuels contracts, with the objective of reestablishing Synfuels as a self-standing and competitive business in the petroleum market. Once this is done, there will be genuinely competitive market pressure from the major bulk gas users—Electricorp, Petrocorp (through Petralgas and NGC), and the revamped Synfuels—to ensure that gas allocations are economically efficient.

Electricity Industry

The electricity industry represents a completely different set of restructuring problems. In essence, the structure of the industry at the bulk level is now set. This structure is dominated by Electricorp, and it is almost inevitable that its effective monopoly will continue. There is potential for competition to occur in generation and bulk supply from a number of different sources. These sources include private firms, other SOEs, and local supply authorities. Whether competition will actually occur remains to be seen. Whether it matters that competition occurs is, in many respects, an even more interesting question. There is some logic to the view that an efficient monopoly may lead to a more economically efficient market than would be achieved through extensive competition.

From a restructuring point of view, there are two remaining areas of concern. The first is whether the internal structure of Electricorp will be effective in allowing competitive generation to develop. Electricorp has recognized its dominant market position by deliberately separating its component businesses. Of central importance is the Transmission Division, which is expected to make transmission services available to all comers, including Electricorp's production division, on a competitively neutral and transparent basis. In effect, Electricorp has opted for self-regulation in order to forestall pressure for imposed regulation.

The other area of concern is the distribution industry. Electricity distribution in New Zealand has traditionally been a function of local government. Thus, a structure consisting of some 60 authorities, some operating as municipal electricity departments of city councils and others as power boards, has matured. These authorities have been politically rather than commercially driven, and the power boards, in particular, have seen themselves primarily as consumer co-operatives. It is likely that deregulation of this industry, as presently structured, would have almost no positive impact and could even be counterproductive.

The key to improved performance is restructuring so as to implant more commercial ethics into the business end of electricity distribution. Politically, however, restructuring is fraught with difficulties. Restructuring policy has been a source of friction between central and local governments for several years; thus progress has been very slow. However, pressure for more sweeping change is mounting, and it is expected that this will come to a head before 1990.

DEREGULATION

The second leg in the economic reform of the energy sector is deregulation. However, perhaps more than in any other sector, deregulation has had to go hand in hand with restructuring. In many respects, it can be argued that deregulation without restructuring would be counterproductive, if not harmful. The reasons for this lie in the particular characteristics of the energy sector, which are in many respects structurally biased against competitively efficient operation. These characteristics include the following:

- The status of many energy products as essential commodities that the consumer has no choice but to use
- The utility nature of the business—the fact that it is supplying a basic input to other activities rather than being an end in itself
- The long-term nature and capital intensiveness of the energy supply business; the impact of decisions last many years, and sunk costs have a dominant influence over sector operations

- The limited number of large individual participants in the energy market, particularly at the bulk supply level.

The following discussion looks primarily at the deregulation of energy markets, with emphasis on electricity, oil, and gas. Since the geothermal and coal markets are already substantially unregulated, they receive only passing mention.

Electricity

In the area of generation and bulk supply, the electricity market will be completely free of economic regulation by early 1988. There will be no price control and no licensing or other controls over generation or transmission. In principle, this could open up the market to widespread competition; in practice, the market is likely to be totally dominated by Electricorp. Therefore, sector performance will depend very much on how Electricorp handles its responsibilities and exercises its wide freedom.

At the local distribution level, there is, in principle, a commitment to deregulation, but little in the way of specifics. Electricity distribution is a natural monopoly and presents major problems in achieving effective deregulation. As indicated earlier, the real priority here is restructuring, and this is the direction in which the ministry will be urging action.

Oil

The downstream oil products market is one of the few energy markets that look like a prime prospect for effective deregulation. There are plenty of participants in the market, products are potentially available from several sources, and it is an easy industry to gain access to—there seems to be a gasoline station on every second corner in New Zealand. For these reasons, the government is firmly committed in principle to deregulation and some initial moves have occurred (for instance, gasoline is no longer under minimum price control). However, both gasoline and diesel remain under maximum price control.

As discussed earlier, the problem centers on the refinery and the need to restructure its operations so as to achieve a genuine freeing up of the market at the wholesale level. If current arrangements continue, deregulation would simply give the existing oil wholesalers effective control over retailing while avoiding any competitive pressure at the wholesale level.

Natural Gas

Natural gas is another area where effective deregulation is dependent on restructuring. At this stage, gas remains under price control, although these controls will cease if not specifically reapplied at the end of 1987. From a regulatory point of view, the key concerns are the gas transmission monopoly enjoyed by

Petrocorp and the strong market position of the local gas utilities. In this sense, the deregulation of the gas industry has much in common with electricity deregulation, and there is a strong similarity between the positions of Petrocorp and Electricorp in their respective markets.

Geothermal

In discussing deregulation of energy markets, geothermal should be mentioned because it presents rather special problems. Geothermal markets are not explicitly regulated at present, aside from the basic licensing of the resource. The question in this case is whether they should be regulated. The key problem is that geothermal is highly location-specific because of the high cost of transport, and this limits the number of viable customers. Moreover, the principal customer is the electricity generation market, in which Electricorp has the overwhelmingly dominant position. Thus, the critical issue is whether the activities of Electricorp in the geothermal market should be constrained to open up the market to greater competition.

ELECTRICITY INDUSTRY: SOME ANALYTICAL ISSUES IN STRUCTURAL AND REGULATORY AREAS

Background

The electricity industry has played an important role in the growth of the New Zealand economy. It has been, and is likely to be, an important source of energy for the residential, commercial, and industrial sectors. In New Zealand, as in most Western developed countries, the industry has been subjected to varying degrees of government regulation.

New Zealand has a population of about 3.2 million, distributed primarily between two islands. North Island has 2.4 million people, and of these, about 800,000 are located in the vicinity of the main center of population, Auckland. Electricity is reticulated over some 266,000 square kilometers (nearly all the land area), and 99.7 percent of the population is connected to the supply. Table 6.1 gives some idea of the size and configuration of the electricity system in New Zealand.

As can be seen from Table 6.1, nearly 70 percent of electricity generation in 1985–1986 was by hydroelectric plant. Of this, nearly 70 percent was generated in South Island. With most of the demand occurring in North Island, there was a net transfer of power from South Island to North Island through the 500 kv DC link connecting the two islands.

The history of the electricity industry has been a mixed one. In the 1960s and early 1970s, the industry enjoyed steady growth, and electricity price increases were modest in relation to price increases of other goods and services. From the mid–1970s, the picture changed quite significantly; only a modest growth was

Table 6.1
Electricity Division Generating Plan (data for year ending March 31, 1986)

Type of plant	Number of stations	Installed capacity (Mw)	Generation (Gwh)
Hydro	30	4,349	18,687
Coal	2	1,118	3,143
Oil/gas	5	1,513	3,047
Geothermal	1	146	1,109
Totals	38	7,126	25,986

Source: Electricity Division, Ministry of Energy, Wellington, New Zealand, 1987.

experienced, resulting in substantial excess capacity and the consequent misallocation of resources. Figures 6.2 and 6.3 give movements in electricity prices and consumption over the 1961–1986 period. (Figure 6.2 also indicates future prices to the year 2005 based on the present trend.)

Unlike some other parts of the energy sector, the electricity subsector is dominated by agencies in the public sector. Generation, transmission, and distribution of electricity are predominantly undertaken by central and local government agencies. This contrasts with the organization and control of other fuel sectors, which are either privately owned and operated or have a combination of private-sector and government interests. Before April 1, 1987, the Electricity Division of the Ministry of Energy (now established as the Electricity Corporation of New Zealand) was the dominant agency in the industry at the generation and transmission level. Approximately 96 percent of electricity consumed was generated by the Electricity Division, the remainder being produced by local electrical supply authorities and a very small amount by privately owned plants. Distribution of electricity at the retail level is undertaken by 60 local supply authorities. There are two types of electrical supply authorities: territorial authorities and power boards. There are 38 electric power boards, which have no other function than to supply electricity to consumers and related activities, and 22 territorial authorities, either municipal or county councils, for which electricity supply is only one of a range of functions.

Before the recent regulatory and structural changes were initiated and partly implemented, generation, transmission, and distribution of electricity was closely regulated. The Electricity Act of 1968 is the primary act governing the conduct of agencies involved in these activities, although there are several other acts dealing with specific operational aspects of the industry. Under the Electricity Act, a person or body wishing to generate, transmit, or distribute electricity needs a license to do so. The price charged by the Electricity Division at the bulk level is governed by provisions of the act and generally subject to approval

Figure 6.2
Electricity Prices (in real March 1986 $)

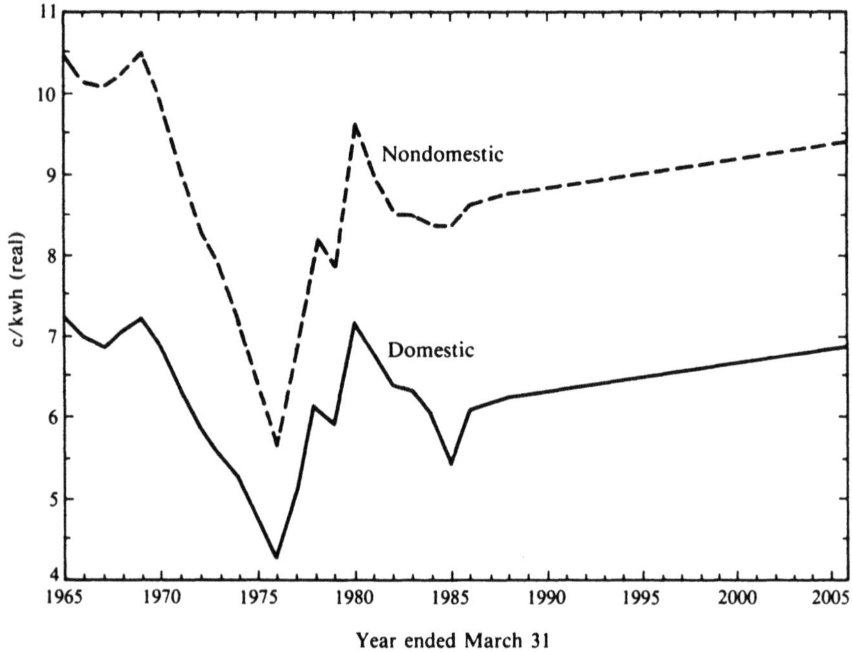

Source: Electricity Division, Ministry of Energy, Wellington, New Zealand, 1987.

by the Ministers of Energy and Finance. Similarly, the central government closely monitors the prices charged by local supply authorities at the retail level. (Most of these regulations have been repealed or are being removed.)

One of the major problems in the electricity industry is the estimation of future demand. Although electricity demand forecasting in New Zealand has been reasonably accurate in the recent past, in the 1970s the slowing in electricity demand was not foreseen in New Zealand as well as in many developed nations. As a result, excess capacity in the electricity system occurred. Two significant forecasting uncertainties—the electricity demands of major industrial projects and future energy prices—continue to cause significant problems for the forecasting process.

Before the establishment of the Electricity Corporation, estimates of future electricity demands of major industrial projects were done by the Ministry of Energy after consultation with existing and potential consumers and appropriate local distribution agencies. There were no incentives in the existing system necessitating large industrial consumers (in fact, any consumers) to accurately indicate their requirements to either the ministry or their local distribution agen-

Figure 6.3
Electricity Consumption, Nondomestic and Domestic Sectors

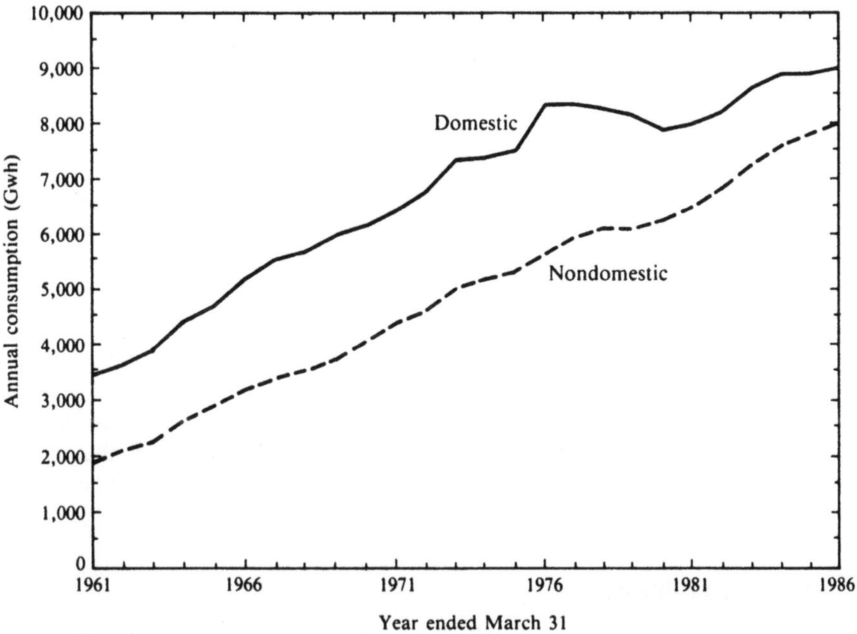

Source: Electricity Division, Ministry of Energy, Wellington, New Zealand, 1987.

cies. In fact, evidence suggests that many industries do not plan very far ahead; for many, the planning horizon is five years or less.

To alleviate some uncertainty with respect to future electricity demands of major projects, some commentators have suggested that potentially large electricity consumers should have to negotiate a commitment for energy supply as a part of project development. It is argued that such a commitment would protect the interests of all parties by reducing the risk of building surplus electricity supply capacity and ensuring supply availability.

Uncertainty over future electricity prices is another factor responsible for overestimation or underestimation of future electricity demand. Future pricing is a major uncertainty for the electricity industry. Several attempts to introduce a degree of stability for both future prices and pricing formulas have been less than successful. Past New Zealand governments have used electricity pricing to serve their own overall political and economic objectives, causing some confusion in the industry. The new Electricity Corporation has issued a major discussion paper on electricity pricing that seeks the views of consumers and others on major parameters affecting electricity pricing.

Some argue that the other major problem that faced the industry was its structure and the overall regulatory environment in which it operated. According to the critics of the past regulatory environment, electricity has been one of the most highly protected and regulated industries in New Zealand. The Electricity Division was, they argue, highly protected by licensing procedures required for hydro generation, transmission, and public sales of electricity; prices were set by the Ministers of Energy and Finance. The local authorities and power boards had the exclusive right to supply certain regions and districts, but only on a nondiscriminating basis under a set of safety and quality regulations. In addition, they were required to purchase their electricity from the Electricity Division. Their prices were not specifically regulated, but their pricing behavior was subject to industry-specific controls. Both the Electricity Division and the regional supply authorities have had wide-ranging powers with respect to land access rights, and have special privileges with respect to water rights and other planning inputs. Finally, all licenses and consents to produce, transmit, or retail electricity were subject to various conditions imposed by the government. The implicated effects of the New Zealand electricity market structure compose the basis that deregulation proponents use to support their position. These are discussed below.

Current Problems

In line with the overall economic philosophy of the government to deregulate the New Zealand economy as far as practically feasible, some major changes have been made regarding the regulatory aspects of the electricity industry. This section presents some of the issues leading to these recent changes. Many analysts have raised questions regarding the costs of government regulation and intervention in the electricity industry. These, they suggest, include distortions to resource allocation, reductions in the incentives to innovate and bear risks, distortions of consumption patterns, and the high administrative and compliance costs of regulation.

Proponents of a complete (or nearly complete) deregulation of the New Zealand electricity industry argue that the present controls in the industry are based on a philosophy of state control that is no longer appropriate and has resulted in an inefficient electricity industry. It is further argued by deregulation proponents that the present regulatory environment has had numerous undesirable effects, particularly the following:

- Inefficiencies in investment decision making, a result of statutory monopoly rights, are demonstrated by the past record of cost overruns of some power projects in the electricity sector. The obligation to supply electricity exacerbates this problem, placing an emphasis on maintaining adequate supplies in a period of rapid demand growth, rather than on cost and competition. Inefficiencies also exist at the retail level, where supply authorities are franchised.

- Inefficient pricing is an additional result of the monopoly environment at both the national and the local level. This is reflected in the pricing policies for wholesale tariffs, which do not cover costs, and the cross subsidization that occurs at both the national and the local levels. In the latter case, international statistics indicate that of the 21 members of the International Energy Agency (IEA), New Zealand is the only country where the average domestic tariffs are below industrial tariffs (excluding direct sales by the Electricity Division). On the average in IEA countries, domestic tariffs are around 75 percent above industrial tariffs.
- The controls reduce the incentives for innovation in the industry and suppress the need to adapt to the changing requirements of customers. This is most readily seen in the supply-side attitude of the Electricity Division, which has taken little account of more innovative and less mainstream alternatives to state supply.

In response to these problems, some of the changes proposed by deregulation proponents were the following:

- Licensing for electricity production, transmission, and distribution should be removed.
- Electricity should not be subject to price control.
- The state and local authority trading enterprises should not be subject to special requirements or receive special advantages vis-à-vis private-sector trading activities (for instance, state and local authorities should be subject to income tax in the same manner as private-sector trading activities).

These changes, as argued by deregulation proponents, would result in a more competitive and efficient electricity sector. In particular, it is suggested that the proposed changes will have the following effects:

- Competition will increase within the industry; consequently there will be increasing pressure on the management of the Electricity Division and the supply authorities to behave as cost minimizers. Lower average prices are expected from a more competitive and efficient electricity sector.
- The industry is likely to be more responsive to the individual and changing needs of customers by allowing contract negotiations between actors to suit their particular needs as well as by providing a wider choice of services at the retail level.

To ensure these results, the proponents of deregulation also proposed a restructuring of the industry. Nine alternative industry structures, each a variation of horizontal and/or vertical disintegration, were suggested. Thus, given an alternative structure, deregulation is expected to bring about a more efficient industry.

The Generation and Transmission Markets

Institutional Changes in April 1987. In May 1986, the government announced its intention to establish an SOE to manage the generation and transmission of electricity. The rationale was that, in the past, the industries had been run with a mixture of commercial and social objectives, and that this had led to poor performance and made accountability difficult. A set of principles was enunciated for the new enterprise:

- Responsibility for noncommercial functions will be separated from trading functions.
- Managers of the SOE will be given a principal objective of running the corporation as a successful business enterprise.
- Managers will be given responsibility for decisions on use of inputs, pricing, and marketing of their output within the performance objectives agreed upon with the ministers so that managers can be held accountable for the results.
- The advantages and disadvantages of the SOE in relation to its private-sector counterpart, including unnecessary barriers to competition, will be removed so that commercial criteria will provide a fair assessment of managerial performance; hence the competitive neutrality of the corporation.
- The SOE will be reconstituted under the guidance of boards generally comprising members appointed from the private sector.

With the passage of the State-owned Enterprises Act in December 1986, nine new SOEs came into being in April 1987, established as limited liability companies under the New Zealand Companies Act of 1955. The Electricity Corporation was one of them. This new corporation has taken over essentially all generation and transmission assets previously managed by the Electricity Division of the Ministry of Energy. The ministry, as far as the electricity industry is concerned, is left with the advisory and regulatory role.

The Electricity Corporation is a public corporation with two shareholders: the Minister of Finance and the Minister of Energy. No other shareholders are allowed. There are, however, provisions for the issuance of equity bonds that will have no attendant voting rights. As opposed to equity, debt for the company will initially all be in the form of government loans, but the intention is that, in due course, this will be refinanced by the corporation's raising loans in the private-sector loan market.

It is possible that the government may wish the Electricity Corporation to undertake certain activities that would not be commercially worthwhile, such as supplying electricity to remote areas. Where this is the case, the State-owned Enterprises Act requires that the government enter into an agreement with the corporation that will provide for the government to pay for such activities. At present, no governmental payments of this nature are contemplated.

The statement of corporate intent of the Electricity Corporation will provide an important component in the mechanism for the shareholding ministers to hold the directors of the corporation accountable. By the State-owned Enterprises Act, the board of directors is required, at the beginning of each financial year, to furnish a statement containing the following for the three upcoming years:

- The objectives of the group
- The nature and scope of the activities to be undertaken
- The ratio of consolidated shareholders' funds to total assets, and definitions of those terms
- The accounting policies
- The performance targets and other measures by which the performance of the group may be judged in relation to its objectives
- An estimate of the amount or proportion of accumulated profits and capital reserves that is intended to be distributed to the Crown (the government)
- The kind of information to be provided to the shareholding ministers by the state enterprise during the course of those financial years, including the information to be included in each semiannual report
- The procedures to be followed before any member of the group subscribes for, purchases, or otherwise acquires shares in any company or other organization
- Any activities for which the board seeks compensation from the Crown (whether or not the Crown has agreed to provide such compensation)
- The board's estimate of the commercial value of the Crown's investment in the group, and the manner and times at which this value is to be reassessed
- Such other matters as are agreed on by the shareholding ministers and the board.

The corporation will be required to produce reports on its operations semiannually, and annual financial statements showing profit and loss and changes of financial position.

The corporation will be expected to pay dividends to the government, and the shareholding ministers may, under the legislation, determine the amount payable.

Valuation of the Electricity Corporation's Assets. Given that the corporation is established as a commercial entity, it is important that an appropriate valuation be given to its assets on the transfer of these assets from the government to the corporation. While the new Electricity Corporation has been managing assets transferred to it since April 1987, at the time of this writing, the sale price of the assets had not been determined.

Officials from the New Zealand Treasury, acting on behalf of the Crown, have been negotiating in closed sessions with the board of the Electricity Corporation on the sale price. The government's position is that the assets should be transferred at market value, because if the price is too low, taxpayers will be shortchanged. Also, such a price would make it relatively easy for the corporation to meet rate-of-return targets, and there would be the possibility that the corporation could distort the national economy by attracting resources from elsewhere, with a resulting loss in allocative efficiency. On the other hand, with the price set too high, the corporation would find it difficult to perform satisfactorily, electricity prices could be forced up (to the disadvantage of consumers), and future taxpayers, as shareholders in the corporation, may not receive adequate returns.

The government's approach to the valuation has been to base it on the net present value of future earnings of the assets. This has been done by relating forecasts of electricity prices to estimated marginal costs and forecasting demand. A number of pricing scenarios have been examined. The discount rate has been chosen using the capital asset pricing model.

Economic Monitoring of the Electricity Corporation. Some of the mechanisms for holding the corporation accountable were outlined earlier. However, these mechanisms, including the use of the statement of corporate intent, are essentially the means by which the shareholders can monitor the performance of their investment, where the shareholders' primary motive must be taken as profit maximization.

As the corporation will generate nearly all the electric power in the country and be the owner of the transmission grid, concern over its taking advantage of its virtual monopoly position in the market has been expressed. However, the government's stance has been one of minimal regulation and control, and of reliance on the Commerce Act of 1986. This act provides that no organization having a dominant position in a market shall use its position to restrict the entry of others; it also contains provisions for invoking price controls. At present, there is no inclination to subject the Electricity Corporation to any kind of price control.

Essentially, it is still unclear how economic/regulatory monitoring will develop. If competition develops, the need for regulatory control may be minimized. It is, however, generally recognized that the generation system the corporation has inherited is overbuilt. This limits the possibility of a new privately owned generating plant's coming on the scene in the near future. A possibility might be to require the corporation to divest itself of some of its plant. If competitive generation is to become a reality, it may be necessary to carefully consider the access that private generation will have to the transmission grid. The board of the corporation has indicated that it will provide access to the grid on an equal basis for its own production and for that of other parties. The Commerce Act provides some safeguards here, but it may later be necessary to have additional provisions to support equal access. As opposed to specific legislative provisions, another approach might be to use voluntary agreements between the government and the corporation. Publication of such agreements could help to ensure the desired effects.

The Distribution Market

The Direction of Change. So far, the distribution market has not undergone the same degree of change as that experienced by the generation and transmission markets, although consideration is being given to whether the structural arrangements of the past are the most appropriate for the future.

The population of New Zealand has grown rapidly in recent decades. In 1950, the number of connections to the electricity supply was about one-third its current level. To some extent, the structure and ethos of the distribution industry can be seen as having been geared for growth and extension of the network.

Critics of the electrical supply authorities argue that the authorities do not have commercial objectives and are not operating in an efficient manner based on sound business principles. Evidence suggests that the authorities' pricing policies favor residential consumers at the expense of commercial and industrial consumers. This, the critics suggest, results in substantial misallocation of resources and consequent inefficiency. They also argue that this is an abuse of the monopoly position enjoyed by the authorities, and is a direct consequence of the fact that members of the boards of supply authorities are elected by the residential consumers.

Thus, the viewpoint in the central government is now developing along the line that increased importance should be attached to ensuring that the electricity distribution industry operates efficiently. As happened with the generation and transmission side of the industry, consideration is being given to the suitability of applying the limited-liability company model to local supply authorities. The basic idea, as it was with setting up the Electricity Corporation, is to have organizations with a commercially oriented structure, clearly stated financial objectives, and management accountable, through a board of directors, to shareholders.

Issues for the Distribution Industry. In the past, there have been various, if implicit, cross subsidizations operated by local supply authorities. Residential consumers have tended to be subsidized by commercial and industrial consumers, and rural consumers have been subsidized by consumers in urban areas. The question arises as to whether such cross-subsidization should continue and, if it should, whether it is compatible with the company model. It seems possible that the shareholding authority could make agreements with the board of directors for the provision of noncommercial activities, following an approach similar to that legislated with the SOEs, including the Electricity Corporation.

There are a number of legal constraints placed on the existing supply authorities:

- They have franchised supply areas.
- There is an obligation to meet all requests for supply.
- There must be nondifferentiation of tariffs within consumer classes.
- There are financial regulations limiting, for example, the terms under which loans can be raised.

Consistent with the general spirit of deregulation within government, it seems likely that many of these constraints will be removed. However, more detailed consideration of these issues is necessary.

Another issue is whether the Electricity Corporation should be allowed to set up a vertically integrated structure by entering the distribution market. It already directly supplies a number of large industrial consumers, such as the aluminum smelter, but it is not involved in any areas of general reticulation. Currently, the corporation has not entered the distribution market, pending the examination of this issue by the government.

The Future of the Energy Sector

With the setting up of the Electricity Corporation and the removal of economic regulations, there have been substantial structural and regulatory changes in the generation and transmission markets of the industry. Future development will be very dependent on the performance of the corporation, both commercially and in its perceived influence on the rest of the economy. Similar structural and regulatory changes are likely in the oil and gas sectors in the near future.

It is clear that the power of the central government to affect the detailed decision making of the energy sector is now substantially reduced. However, it is not clear whether this will produce some kind of reaction from the government. For example, governments, particularly newly elected governments, may be tempted to make changes in the board of directors in an attempt to steer the corporation according to their wishes. Despite this possibility, it seems most unlikely that the broad direction of change—that is, away from integrated planning at the national level and toward commercially oriented, market-based decision making—will be reversed.

There have already been some moves to introduce partial private ownership in hitherto state-owned assets in New Zealand. This has been done through the issue of shares in the Bank of New Zealand and in the Petroleum Corporation. While the government may be reluctant to make changes too quickly in the newly formed Electricity Corporation, it seems quite possible that there could, in due course, be moves to introduce an element of private ownership.

Although a few local electrical supply authorities are planning transformation into companies, there has been little significant change in the structure of the distribution industry so far. In the future, further moves to set up companies are possible, and, given the existing large number of small authorities, such moves could precipitate a degree of rationalization and amalgamation of authorities. The power of the central government to influence the locally based distribution industry is somewhat less than its power over the agencies of the central government, such as the Ministry of Energy. As a result, the pace of change in the distribution industry will be dependent on local predilections, and may well be much slower than has been the case for generation and transmission. On the other hand, if area franchising is dispensed with, it is possible that a few aggressive authorities could endeavor to capture customers across their historical boundaries, thereby producing a domino effect speeding up the process of change.

It is interesting to speculate on how the industry at the operational level will respond to the structural changes. What sort of development strategy will the Electricity Corporation adopt? As indicated earlier, there is at present generating plant excess capacity, and this means little need for major capital expenditure on power stations until around 2000. It is, however, possible that some work will be done to increase the capacity of the DC link between the two islands before then. Generally, it seems that the establishment of the corporation will have the effect of deferring major new developments. This may, in part, be viewed as the corporation's having a somewhat risk-averse strategy. With the great uncertainty over future demand for electricity, it may be prudent for the corporation to sit on its current assets and wait for demand growth to occur before making new investments. In the meantime, if unexpectedly high demand does occur, this could be reduced by raising prices or by planning investments with a shorter construction period. This approach could provide a convenient way of financing investments.

A major consideration for the industry will clearly be the future pattern in the growth of demand. Institutional and structural changes in the New Zealand economy beyond the electricity industry, including the establishment of state-owned enterprises such as the Forestry Corporation, could conceivably have some impact on the price/demand relationship for electricity. A key factor in determining demand will be the pricing of electricity.

The Electricity Corporation is investigating ways of making the wholesale pricing system more responsive to the costs of supply. Time-of-use and real-time pricing, as well as the introduction of regional prices, are being considered along with possibilities of options such as operating a spot market. This raises a degree of uncertainty for the local supply authorities in terms of their electricity purchase costs.

The broad thrust of the changes taking place in the New Zealand electricity industry is to make for an efficient industry that is much more responsive to changing costs, technologies, and people's needs. Thus, in general, the industry is being opened up to the uncertainties of the world.

CONCLUSIONS

Since 1984, the energy sector in New Zealand has undergone the most comprehensive program of structural and regulatory changes in this century. The integrated energy planning approach has been replaced by a greater reliance on the market as the best means of driving decision making. This, in turn, has dramatically changed the emphasis of government's role in the management of the energy sector.

The new regime dictates that decisions on these issues be made primarily on commercial grounds by corporation boards of directors, rather than by ministers of the Crown. However, there will continue to be a strong role for government in ensuring that long-term strategic factors and broader social issues are taken into account in the decision-making process. However, the means for intervention will be different. Rather than intervening in specific decisions, the means will be through the process of creating an appropriate environment for commercial decision making. The challenge for government and for policy advisers is to ensure that the new environment does indeed harness commercial decision making to the national good, so that commercial and national interests are in harmony and not in conflict.

SEVEN

Impacts of Deregulation on the Australian Petroleum Industry

SHARON HOFFMAN

INTRODUCTION

Deregulation and privatization have become critical issues in many countries of the Asia-Pacific region. Regulations on petroleum production and marketing were originally created to encourage the development of domestic production and refining, as well as to protect the fledgling industry from world market pressures. Since the genesis of Australia's petroleum industry in the 1960s, however, that fledgling industry has grown not only in size but in sophistication as well. Given its newfound strength, coupled with the reality that Australia cannot forever maintain its high level of petroleum self-sufficiency, Australians have reflected closely on their past ideology.

The resultant revelation materialized, as of January 1988, into a deregulated petroleum market. Australia does not stand alone in the efforts to integrate its petroleum sectors into the world market. Efforts are under way in New Zealand and Japan, and the Philippines and Thailand are also considering loosening their grip on their petroleum industries. Though Japanese deregulation is clearly the most significant, its slow pace means the impact will not be felt in the short term. In contrast, the rapid changes in Australia will provide lessons to other governments in the region in the importance of deregulation.

This paper draws heavily on S. Hoffman and F. Fesharaki, *Deregulating Australia's Oil: The Implications for Global and Regional Markets* (London: Economist Intelligence Unit, 1987).

THE RATIONALE FOR DEREGULATION

While specific reasons vary as to why countries decide to deregulate their petroleum industries, it is hoped in all cases that the move into the world market will bring about rationalization of the domestic petroleum sector as well as the establishment of valuable marketing networks that will ensure the satisfaction of petroleum needs in the most efficient manner possible. For Australia, one of the specific motivations for deregulating is its reliance on a declining reserve base; probable and proven reserve estimates of crude oil and condensate total more than 2.8 billion barrels, of which 80 percent is in the Bass Strait of the Gippsland Basin. At a consumption rate of 580,000 barrels per day (b/d), the reserve base has only 12 years of life remaining. Other important basins include the Timor Sea, the Northwest Shelf, and the Cooper-Eromanga, Canning, Amadeus, and Jurat Basins; however, none is so critical to Australia's energy future as Gippsland.

The future production profile for Gippsland is one of decline; though the severity of this decline may be ameliorated by the recent tax concessions (to be discussed in detail later). The turn of the century will mark the time when Australia, barring new discoveries, will be very much in need of imported petroleum. At that time, there will be very little domestic production to regulate.

While Australia may see its future as deficient in reserve base, it is in no way lacking in domestic expertise in refining, production, and marketing. Multinationals are a major force—Shell, BP, Esso, Caltex, and Mobil are all active in refinery and/or exploration. Esso and Broken Hill Proprietary are by far the largest oil producers in Australia, as they alone control the Gippsland Basin. The retail market, however, is dominated by Shell, BP, Esso, Caltex, Mobil, and Ampol. Shell is by far the strongest force, particularly in the eastern half of Australia, while BP dominates in Western Australia. Esso and Ampol control much smaller shares. Ampol is distinguished by the fact that it is an indigenous company, giving it considerable political clout.

The ownership of refineries is listed in Table 7.1. Australian refiners, with over 790,000 b/d of relatively sophisticated refining capacity, can adapt to deregulation and in many cases have been screaming for it because cheaper imported crudes may yield a product slate in line with Australia's needs.

THE HISTORY OF REGULATION

The Australian oil industry, which has existed only since the early 1960s, has been regulated since 1965, when all indigenous crude oil was marketed under the crude allocation scheme. This was fundamentally a voluntary arrangement between producers and refiner/marketers administered by the Commonwealth government. Under this allocation scheme, producers were guaranteed a market for their product through provision of a mechanism for equitably allocating indigenous crude oil to refining companies. Since January 1985, however, the

Table 7.1

Australian Refining Capacity as of January 2, 1987 (barrels per stream-day)

Company	Location	Primary capacity	Vacuum distillation	Catalytic cracking	Re-forming	Hydro-treater	Polymerization alkylation isomerization	Desulfur-ization
Ampol Refineries	Lytton, Qld.	65,000	26,000	26,000	13,000	—	5,800 I 2,500 P 4,000 A	—
Caltex Refining Co. Pty. Ltd.	Kurnell, NSW	146,000 to 154,000	45,000	43,000	13,500 14,000	16,000	3,000 P 2,900 A	—
BP Refining Pty. Ltd.	Bulwer Island, Qld.	48,000	15,000	14,500	9,000	5,700	1,800 A	8,500
BP Refining Pty. Ltd.	Kwinana, WA	120,000	20,000	25,000	16,000	16,000	2,750 A	8,000
Petroleum Refineries Australia Pty. Ltd.	Altona, Vic.	108,000	18,500 40,000	30,000	31,000	44,000 NH 12,500 DH 12,300 GT	2,900 A	—
Petroleum Refineries Australia Pty. Ltd.	Port Stanvac, SA	72,000	3,000	—	18,000	—	—	46,000
Shell Refining	Geelong, Vic.	110,000 to 132,000	na	25,750	—	57,000	9,250 I 4,250 A 900 P	11,500
Shell Refining	Clyde, NSW	64,000 to 91,000	32,750	28,600	—	36,500	7,500 A 3,600 A 900 P	—
Eromanga Oil Refinery	Eromanga, SW Qld.	1,500	—	—	—	—	—	—
Total		734,500 to 791,500	200,250	192,850	114,500	200,000	52,050	74,000

I = Isomerization. P = Polymerization. A = Alkylation. NH = Naphtha hydrotreater. DH = Distillate hydrotreater. GT = Gasoline treater. — = Not available.

Source: Australian Institute of Petroleum, *The Australian Petroleum Industry in 1987*, January 1987.

allocation system has been altered and has evolved into a free market system through a partial allocation scheme. This move was deemed appropriate for a variety of reasons. There was, in 1983, a technical surplus of indigenous oil that appeared to be consistent for several years to come. It was also known that the surplus could not last, and that in fact production would eventually decline, forcing Australia to operate in a free market. Thus, the partial allocation system was a transitional mechanism.

The main features of the partial allocation scheme were the following:

- All production from small fields producing up to 50,000 b/d (except for extremely remote locations) was sold through allocating arrangements at the import parity price (IPP), which was based on the nominal cost of imported Saudi light crude oil.
- Most production from the Bass Strait was sold through allocation. For 1985, 1986, and 1987, the refiners' allocation of Bass Strait crude oil was 350,000 b/d.
- At production levels higher than 350,000 b/d, surplus crude was available for export or for sale by negotiation to Australian refineries and marketers who had already fulfilled their lifting obligations.
- All crude sold under allocation was sold at the IPP.

It was also the intent of the Commonwealth to use partial allocation as a temporary measure; therefore it was formally initiated for 1985, 1986, and 1987. While the IPP was useful for price standardization, it was also a convenient measure by which excise could be levied. In 1985 the IPP was reviewed every two months and the price was adjusted if movements in the relevant parameters caused a price change of A$1 or more per barrel. In 1986 and 1987, the IPP was reviewed monthly because world crude oil prices were changing rapidly. Now, in the deregulated market, the IPP is no longer in use. In its place is a tax reference price, which is a volume-weighted average of BHP and Esso realized sales prices (VOLWARE price).

TAXATION UNDER REGULATION

While the Australian government regulated the petroleum industry, it also relied heavily on it for tax revenue. Like most governments, it views indigenous petroleum as a national resource, so it devised an extensive tax system to reallocate some of the profits from oil production to serve Australian national interests. The tax vehicles by which this is accomplished are the crude oil excise, resource rent tax, resource rent royalty, royalties, and company tax.

The Crude Oil Excise

The crude oil excise has always been a production-related tax calculated as a percentage of the IPP. The IPP was utilized as the basis for tax calculation in

Table 7.2
Producer Return/Excise Component of Gippsland Import Parity Price

Annual production range per field (thousand barrels)	Excise (% of Gippsland IPP) "Old" oil	"Intermediate" oil	"New" oil	Pretax/royalty production return (% of Gippsland IPP) "Old" oil	"Intermediate" oil	"New" oil
0–315	0	0	0	100	100	100
315–629	5	0	0	95	100	100
629–1,259	15	0	0	85	100	100
1,259–1,888	20	0	0	80	100	100
1,888–2,517	40	15	0	60	85	100
2,517–3,146	70	30	0	30	70	100
3,146–3,776	79[a]	50	10	20	50	90
3,776–4,405	79[b]	55	20	20	45	80
4,405–5,034	79[b]	55	30	20	45	70
5,034+	79[b]	55	35	20	45	65

a. Originally 80 percent.

b. Originally 87 percent. The 79 percent rate will decline to 77 percent in 1988–1989 and 75 percent in subsequent years.

Source: BHP Marketing Review, August 1987.

the regulated Australian petroleum market and for the most part, with the notable exception of the period 1972–1977, it was higher than the world market price.

Although the crude oil excise is a production-related tax, it also differentiates by categories of oil. Historically, oil has been classified into three categories for taxation purposes:

1. "Old" oil, which is oil discovered prior to September 18, 1975, except where oil is deemed to meet the requirements of category 3
2. "New" oil, which is oil discovered from pools (discrete accumulations) after September 18, 1975; prior to July 1984, this oil did not carry an excise, so exploration was stimulated
3. "Old" oil fields that had not been developed by October 23, 1984; an "intermediate" scale designed to allow the development of fields that would prove subeconomic if the "old" oil excise was applied.

Table 7.2 illustrates the producer return and the excise components of Gippsland crude prior to deregulation. Small discoveries were fairly well protected, in that the excise increased with the rate of production from the field, yet remained fairly low on new discoveries. From table 7.2, it can be seen that the top rate of taxation for "old" Gippsland is 79 percent. Until July 1987 this top rate had been 80 percent. However, in its attempts to stimulate production, the government has devised a plan whereby this excise on large fields will be reduced to 75 percent by 1989.

Since the IPP will no longer exist because of deregulation, the VOLWARE price will be utilized for taxation purposes. In addition, in June 1987 the government announced that the first 30 million barrels of crude oil produced from new projects in the Bass Strait (and from the Northwest Shelf and offshore fields) would be exempt from excise. The industry has thus gained some concessions, since excise in 1986 was as high as 87 percent on large fields.

This scale of taxation has been strictly for domestically allocated sales; "old" oil sold in the free market is subject to a sliding scale of excise. Still, it was essentially in the "old" producing fields that return on investment occurred and the tax burden was heaviest. Because mature fields in Gippsland were taxed so heavily, producers felt there was a large disincentive for additional exploration. Given Australia's domestic production outlook, exploration is clearly needed.

The Resource Rent Tax

During the legislative session on December 15, 1987, the resource rent tax (RRT) was passed. The legislation provides for the application of RRT to offshore petroleum projects that had not reached the development stage (known as "Greenfields projects") as of July 1, 1984. The RRT replaces both excise and royalty payments on those fields where it applies. Those areas covered by production licenses granted prior to this date and the permit areas from which those licenses were drawn are excluded (specifically, VIC–P–1, and WA–28–P). The Jabiru project, which began crude production in 1986, was already subject to the RRT.

The basic features of the RRT are the following:

- The tax is applicable to individual projects.
- The tax is levied before the company tax, and payments of RRT are a deduction for company tax purposes.
- The threshold rate is set at the long-term bond rate plus 15 percent. Thus, projects are assured a minimum rate of return before taxation is applied.
- The tax rate is 40 percent.
- Exploration expenditures in the permit area for five years preceding the granting of a production license may be compounded forward at the threshold rate for deduction from assessable income.

The Resource Rent Royalty

The resource rent royalty (RRR) is the onshore corollary of RRT. It applies to onshore areas where the state government and the companies prefer the RRR to a combination of excise and royalty payments. To date, the only area where the RRR applies is Barrow Island.

Royalties

As mentioned, the RRT, where applied, replaces the combination of crude excise tax plus royalties. At present, royalty arrangements in offshore areas outside the three-mile territorial sea are under Commonwealth jurisdiction, whereas the states control onshore royalty arrangements and those within the territorial sea and coastal waters. In offshore fields where royalties apply, the basis is the wellhead value agreed upon by the taxing authority and the producer. The tax rate is between 10 and 12.5 percent of the wellhead value. This tax rate is composed of two parts: the basic 10 percent rate for a primary license, the proceeds of which are shared between the Commonwealth and the state on a 40/60 basis, and an override royalty of between 1 and 2.5 percent applied when a secondary license is granted, the proceeds of which go totally to the state. In the Gippsland oil fields, the royalty rate is 12.5 percent. Royalty to the federal government is 4 percent in all cases, with the balance going to the state.

For onshore fields, the royalty rate is 10 percent of wellhead value, which is calculated by deducting from the value of petroleum sold those costs and expenses incurred downstream of the wellhead up to the point of sale. For Barrow Island, the RRR replaces the crude oil excise and state royalty. The RRR payment is shared on a 75/25 basis between the federal and Western Australian governments.

Petroleum taxation in Australia yields a large amount of revenue for the federal government. In 1986, crude oil and product excise levies, LPG excise levies, royalties, and company taxes yielded approximately A$8.1 billion. It is even more interesting to note the percentage of taxes compared with revenues that a company in Australia typically keeps. Since excise alone accounted for 61.9 percent of oil revenues in 1986, it is easy to understand why the oil companies have looked upon the excise as a burden.

CURRENT TAX LAWS

In the spirit of deregulation, the taxation of petroleum has been reexamined. Under the concept that less tax means more investment and production, a new package came into effect as of July 1, 1987, that reflects an agreement between the oil companies and the federal government. The government agreed to reduce the top marginal rate of excise to 75 percent by 1990 for existing Bass Strait production. For new offshore projects in the Bass Strait beyond the territorial sea, there will be no crude oil excise up to a cumulative production of 30 million barrels from the eligible petroleum accumulation relevant to each project, but excluding production that would otherwise have been captured by prior developments. This concession also applies to new projects on the Northwest Shelf.

The same rules apply to onshore projects, except that fields that commenced

Table 7.3
Gippsland Crude Oil Production Forecasts[a]
(thousand barrels per day)

Year	Forecast I	Forecast II	Additions
1987	375	410	35
1988	300	380	80
1989	225	310	85
1990	170	275	105
1991	150	250	100
1995	90	170	80
2000	60	120	60

Forecast I: Production without tax changes.

Forecast II: Production with lower excise tax, resulting in both new developments and upgrading of existing facilities.

a. Data arrived at from measurement of graphs.

Source: Esso/BHP.

production after January 1, 1987, are subject to the "new" oil excise regime on any production in excess of 30 million barrels. Production volumes of less than 30 million barrels carry no excise obligation.

In addition, all LPG produced onshore will be exempt from excise. Offshore and onshore condensate marketed separately from crude oil is also exempt from excise.

THE IMPACT OF TAX CONCESSIONS ON GOVERNMENT REVENUES

What is the government getting for these concessions? After all, it will forgo approximately A$18 million in 1987–1988 and some A$297 million between then and 1993. The government expects a lot for its money. On the exploration side, this package is supposed to guarantee A$600 million in investment in new offshore and onshore exploration over the period 1988–1993. On the development side, some A$400 to A$600 million will be invested in the Bass Strait, including the development of production facilities at six new fields. Another A$300 million is planned to be applied to enhanced recovery from five existing platforms.

In all, this package was created with the intention that more oil would be produced. For example, 1988 production levels at Gippsland are hoped to exceed 380,000 b/d, as opposed to the forecast rate of 300,000 b/d (Table 7.3). Future production will be increased so that the expected steep declines in Gippsland oil production are slowed. Higher production will generate additional excise royalty revenues for Commonwealth and state governments to the tune of A$3.2 billion during 1988 to 1993.

Everyone hopes this package will provide the impetus for the discovery of another Bass Strait, but the inevitable decline in production is imminent and it seems likely that oil imports will rise.

FUTURE OIL PRODUCTION OUTLOOK

It is expected that additional production will result from the more relaxed excise scheme (Table 7.3). Although it will begin slowly, it is forecast that by 1990 producers will be able to bring onstream an additional 100,000 b/d. After 1990 these additions are expected to decline; by 2000 they will amount to only 60,000 b/d. The excise adjustment, giving rise to increased production, is thus a temporary measure to avoid a large import bill.

IMPACTS OF DEREGULATION

The reduction in excise was backed by the entire industry, but the issue of deregulation of the allocation system for domestic refining divides the industry. The following passages detail the specific positions of those in Australia most affected by deregulation: the refiners, the producers, the shippers, and the government.

Refiners

Refiners were generally in favor of deregulation. They found the IPP excessive and felt the government had limited their options by forcing them to buy only domestic crudes. Even the refiners with pipelines connected to Gippsland (Shell's Geelong refinery and Mobil/Esso's Altona) believed they could play the world market more efficiently under deregulation. The Australian refiners had generally invested in upgraded facilities and were capable of cracking fuel oils from heavier, lower-priced crudes—a capacity they could not fully exploit with low-fuel-oil-yielding Gippsland crude. Furthermore, the refiners felt a more flexible mix of crude was needed to produce the refined products demanded by the consumer. Once they were free to choose their crudes, they could operate more efficiently by refining the proper mix. For these reasons, the refiners' opposition to the allocation system and the IPP was strong and nearly unanimous. The exception was Ampol. Ampol's refinery is designed to run on sweet crudes, and the Gippsland diet suited it best. Ampol was concerned that deregulation might result in its refinery not having access to sufficient volumes of crude. Although it is a small-scale refiner, as an indigenous Australian refiner/producer, Ampol has political influence that cannot be ignored. To avoid any disruption of progress toward deregulation, Esso has given strong assurances to Ampol that it will supply the refinery with the needed Gippsland crude at commercial prices. Ampol has apparently found these assurances acceptable and has not pursued the issue.

Large Producers

The large producers—members of the Gippsland Consortium—were cautious but generally supported deregulation. In principle, they wanted less government interference, but they were also concerned about the consequences of giving up assured markets. As an affiliate of the largest oil company in the world, Esso enjoyed many options. It could refine what it needed in its own refinery in Australia or export to affiliate refineries in Japan, Singapore, Thailand, and the United States. Since almost all of Esso's share of Gippsland was absorbed within its own refining network, Esso felt it could handle the deregulation. BHP, on the other hand, has no refining system, and its exports are fully price responsive. A well-run and aggressive Australian company, BHP felt that no matter how prices moved initially, the flexibility obtained by deregulation would be valuable in the long term. BHP plans to use its skills to exploit special areas of opportunity, negotiate agreements at good prices, and begin to transform itself into a truly international oil company. As long as the government could dictate the destination of its crudes, BHP felt it could not reach its maximum marketing capabilities. In the end, Esso and BHP strongly supported deregulation.

Medium and Small Producers

Outside the Bass Strait, there are many producers in Australia with differing opinions on the impact of deregulation. Generally, medium and small producers felt that with deregulation, refiners might refuse to lift their crudes or might offer such low prices that they would be forced out of business. Many of the producers argued passionately against deregulation. Threats to national security were often cited—it was argued that deregulation would result in lower oil production in Australia at a time when Bass Strait output was about to decline, and that this would increase the country's vulnerability through dependence on imports.

For a time it seemed the government would decide on partial deregulation, essentially deregulating the Bass Strait but keeping all other production allocated to domestic refining. This would have necessitated retaining an IPP applicable to one part of Australian output as well as a free market in the Bass Strait. Esso objected strongly, saying that the government should deregulate in full or not at all. In the end, Minister Evans asked for a clarification of what was really in danger. He understood that the producers of more than 5,000 b/d need not suffer from the free market, and he rejected arguments against deregulation by this category of producers, which he called "medium-sized." He was then told that the total of all other small producers together is less than 5,000 b/d. He announced that the government will seek assurances from the refiners that they will buy crude from these small producers at market-related prices. The government felt that the deregulation issue was simply too important to be jeopardized by a group of small producers, so these cosmetic assurances were given. The fact is, many of these small producers may not survive deregulation.

Though refiners may initially try to assist them, in time the market economies will prevail and, as happened in the United States, many small producers may go out of business. To illustrate the dilemma of these producers, for some the transportation cost to the refineries could be as high as U.S.$7 per barrel, compared with about U.S.$1 per barrel from Saudi Arabia. The scale of production is so small that, to make transportation economical, oil has to be produced and then stored for perhaps six months to make larger shipments possible. This in turn requires investments in extra storage tanks. Moreover, as in the United States, shipping within Australia has to take place on Australian-flag ships at inflated prices.

Thus, deregulation is likely to result in some closures. However, some oil producers are considering innovative ways of pooling their oil and transporting it in larger quantities to refineries. This group has received offers of assistance from a number of companies, including BHP. It is important to understand that most small producers in Australia are actually controlled by larger producers. Independent small producers have a total production not exceeding 1,000 b/d collectively and, as such, their role in the future of the Australian petroleum industry is minor. This explains why the Australian government deregulated without being overly concerned about its impact on small producers.

Shippers

With the advent of more efficient ships with smaller crews, the Australian shipping industry has vastly improved in the past few years. This efficiency will be vital in the competitive environment awaiting it, as it has the opportunity to participate in international trade. The government has had concerns for the shipping industry and has sought assurances from refineries and major producers that they will use Australian-flag tankers when it is commercially practicable. Refineries have also given assurances that they will maintain their existing arrangements for Australian-flag shipping. However, to take advantage of new opportunities in export/import trade, the Australian shipping industry will have to prove itself in competition.

The Government

One of the central concerns of the Australian government regarding deregulation is the balance-of-payments outcome. It is not possible to know exactly what the effects will be, but the Department of Primary Industries and Energy believes there will be either a neutral or a positive effect on the balance of payments. The refiners' view is that under deregulation they will have access to lower-priced crudes. Because of this, they expect that a greater quantity of higher-quality crude will be exported at higher prices. Thus, refiners believe there will be a net gain in the balance of payments. Despite all the reshuffling of crude oil, however, the expected decline in Bass Strait production will inevitably translate into greater quantities of imported crude, and thus a higher import bill.

Deregulation and the concurrent excise arrangements made for 1988 will confront the government with changes and adjustments. Central to this will be the effects of deregulation on internal budgetary concerns, especially since the 1985–1986 revenue from taxes on crude oil, natural gas, LPG, and petroleum products was more than A$8.1 billion.

Due to deregulation, the price of crude oil is forecast to decline by between A$1.00 and A$4.00 per barrel. Working assumptions have placed the likely decline at A$2.50 against the IPP, for a drop in revenue of about A$100 million in 1987–1988. It should be noted, however, that additional production from excise concessions will translate into a net benefit in tax revenues despite this price effect. No immediate product excise increase is foreseen to offset the concessions given in excise taxes or the drop in tax revenue due to lowered oil prices. It is unlikely, however, that product prices will not rise in the future.

For the medium term, excise concessions will eventually give rise to more crude oil production, which in turn will provide additional revenues. It seems highly unlikely that the government will forgo needed revenues. As deregulation proceeds, the government will adapt and will use whatever fair means it can to enhance revenue earnings.

THE IMPACT OF DEREGULATION ON EXPORTS AND IMPORTS

Under deregulation, the changing export nature of Gippsland could be of great importance to oil traders. In the relatively thin Asian oil market, an extra 30,000–40,000 b/d of demand for Minas-type crudes could push up the price of Minas in the early part of 1988, thereby negating some of the benefits of the free market for the Australian refiners. Moreover, exporters of lighter crudes—such as Labuan, Miri, Seria, Attaka, and Tapis—may see a weakening of prices as these crudes face competition from the additional light Gippsland crude. Thus the differentials between light and heavy crudes may narrow in the regional crude market. Had this change happened in Rotterdam or the U.S. Gulf Coast, very little price impact would have resulted, but in the Asian market certain price impacts will inevitably be observed.

WILL REGULATION RETURN?

The deregulation of the Australian petroleum industry has come at an unusual time, when Australia is moving from self-sufficiency in petroleum to a position of large-scale dependency on imported oil—perhaps to as much as 50 percent by the early 1990s. Deregulation and lower taxes will result in both a slowing of the rates of decline and more efficient operation of the oil industry in Australia.

The inevitable question is whether deregulation in Australia can be regarded as irreversible. This is doubtful. Although few would want to believe that deregulation might be reversed after all the efforts of the industry and government, we must caution that regulation could return in a few years. Assuming that by the early 1990s Australia is dependent on foreign oil imports for 50 percent of its needs, what would happen if a crisis in the Middle East led to fears of a supply interruption? Would any Australian government be able to withstand political arguments that import vulnerability was so endangering national security that exports should be banned and first priority given to domestic refiners? The question may be academic—it is possible that domestic producers may be able to extract such high prices from refiners that they would not want to export. Nevertheless, such a possibility should be kept in mind, if only to be prepared in case it is necessary to fight the return of regulation.

PART III
Country Case Studies: East Asia

EIGHT

Effects of Oil Price Changes on the Korean Economy

JEONG-SHIK SHIN

INTRODUCTION

The key role of the energy sector in developing countries has received special recognition since the two oil shocks in the 1970s. Korea is an energy-importing, newly industrializing country that tries to achieve high economic growth with minimum energy import bills.

This chapter investigates the impacts on the Korean economy of changing energy prices, including the two oil shocks in the 1970s and the surprising oil price drops in 1986.

The energy-economy model of Korea is presented and used to analyze the impacts of the changing oil prices on economic growth, energy demand, inflation, and the trade balance.

While the empirical results reported here are preliminary, they provide interesting implications for future studies in this area.

TRENDS IN ENERGY CONSUMPTION

Economic Growth and Energy Demand

Table 8.1 provides historical data on the average growth rates of GDP and total energy consumption in Korea for selected years. During 1970–1986, GDP and energy consumption grew at an average rate of 8.4 and 7.3 percent, respectively.

The GDP elasticities[1] of energy consumption differ from period to period, and energy intensity, measured by the energy/GDP ratio, decreases or increases, respectively, as the GDP elasticity is inelastic or elastic. The increase in energy

Table 8.1
Average Annual Rates of Change in GDP and Energy Consumption

Period	GDP[a] (%)	Energy consumption (%)	Energy-GDP elasticity	Energy/GDP (%)
Pre-oil shock 1970–73	9.8	8.3	0.85	-1.4
First shock 1973–74	8.6	3.0	0.35	-5.1
Adjustment 1975–79	10.6	11.9	1.12	1.2
Second shock 1979–80	-3.0	1.5	—	4.7
Adjustment 1981–85	7.6	5.2	0.68	-2.3
Oil price drop 1985–86	11.95	9.15	0.76	-2.5

— = Negative GDP growth.
a. Based on 1980 prices.
Sources: Ministry of Energy and Resources, *Yearbook of Energy Statistics* (Seoul: The Ministry, 1987 and various issues); Bank of Korea, *National Accounts* (Seoul: Bank of Korea, 1987 and various issues).

intensity during 1979–1980 is a result of government-induced investment in energy-intensive heavy industries in the late 1970s.

Changes in the Composition of Fuel Demand

Total consumption of energy in Korea increased from 19.679 million metric tons of oil equivalent in 1970 to 61.065 million in 1986, an average annual growth rate of 7.3 percent. Primary energy consumption in Korea is sensitive to changes in the price of oil.

The share of petroleum in total energy consumption was as high as 62.7 percent just before the second oil shock. Since the second oil shock, reduction of oil dependence has been one of the most important issues facing the Korean government. As a result, there has been massive substitution of nuclear and coal for oil in both the power generation and the industrial sectors. The demand for oil showed rapid growth (11.7 percent) in 1986, reflecting the effects of the sharp decline in the oil price and the high growth of oil demand in the transportation sector. Still, oil dependence has been reduced through the substitution of nuclear and coal for oil in the power generation sector.

IMPACTS OF OIL PRICE CHANGES ON THE ECONOMY

Impacts on Inflation

Oil price changes had a significant impact on the inflation rates in Korea. During the first and second oil shocks, the wholesale price index rose by 41.9 and 38.9 percent, respectively. When the oil price was relatively stable during the period 1975–1979, the index moved at almost the same pace.

Impacts on GDP and Trade

Korea has shown economic growth, but it has not been immune to disturbances in the international oil market. During the second oil shock, its economy experienced negative growth.

Energy import bills increased annually by 244.9 percent during the first oil shock, and by 121.8 percent during the second. Total imports increased rapidly over these two periods, at rates of 64.1 and 37.6 percent, respectively.

During the second oil shock, total exports grew faster than total imports. This mainly reflects the effects of changes in the exchange rate of the won versus the U.S. dollar. During 1979–1980, the won depreciated by about 26 percent against the dollar.

THE ENERGY-ECONOMY MODEL

Among the models developed to analyze the impacts of oil price changes on energy demand and economic growth, I extend the Chern et al. type of model in the Korean situation because of its simplicity and good performance. Data constraints prevent us from using more sophisticated approaches.

The model consists of six structural equations. The first is the weighted domestic energy price index, a function of the imported oil price index. Primary energy consumption is a function of real GDP and the weighted energy price index deflated by the wholesale price index. A partial adjustment model is applied. The GDP equation is an aggregate production function. It determines output as a function of capital, labor and energy.

The inflation rate is measured by changes in the wholesale price index, which is a function of the ratios of money supply to GDP and energy prices. Total exports are a function of GDP and the relative export price index. Total imports are a function of GDP and the imported oil price.

VARIABLE DEFINITIONS AND DATA SOURCES

Variable Definitions

Endogenous variables:

WEP = Weighted energy price index (1980 = 100)

PEC = Primary energy consumption (thousand metric tons oil equivalent)

GDP = Real gross domestic product (billion 1980 won)

WPI = Wholesale price index (1980 = 100)

TX = Total exports (billion won)

TM = Total imports (billion won)

Exogenous variables:

PDO = Imported oil price index (1980 = 100)
Evaluated in won terms
(U.S.$ in FOB × exchange rate [won/U.S.$])

KSI = Capital stock of previous year (billion 1980 won)

LAB = Number of employees (thousand persons)

MM2 = Money supply (M2) (billion won)

RPX = Relative export price index (1980 = 100)
Derived by dividing the unit value of Korean exports (valued in U.S.$) by the unit value of Korean imports from major trading countries (Japan and the United States)

D80 = Dummy variable
1 for the year 1980
0 for other

Data Sources

The time series data from 1970 to 1985 are used for estimation. Data on the weighted energy price index, primary energy consumption, and imported oil price are obtained from the *Yearbook of Energy Statistics* published by the Ministry of Energy and Resources. GDP and money supply data are from *National Accounts* and the *Monthly Bulletin* published by the Bank of Korea. Data on the wholesale price index and the number of employees are from the *Major Statistics of the Korean Economy*, published by the Economic Planning Board. Data on total exports and imports, unit value of exports, unit value of imports from other countries, and exchange rates are from *International Financial Statistics* of the International Monetary Fund. Capital stock data are derived from *National Accounts*.

ESTIMATION PROCEDURE AND RESULTS

In principle, the method of analysis for the system of equations that follows should be nonlinear, three-stage least squares (3SLS). However, the application of nonlinear 3SLS to the present data set did not produce reasonable results. Thus, the results from ordinary least squares are presented below (t-ratios are in parentheses).

(1) $\ln \text{WEP} = 1.023 + 0.786 \ln \text{PDO}$
$(8.79) \quad (25.61)$
$R^2 = 0.98$

(2) $\ln \text{PEC} = -0.379 + 0.578 \ln \text{PEC1} + 0.515 \ln \text{GDP}$
$(-0.77) \quad (3.87) \quad (4.52)$
$-0.121 \ln (\text{WEP/WPI} \cdot 100)$
(-1.70)
$R^2 - 0.99$

(3) $\ln \text{GDP} = -4.164 + 0.176 \ln \text{KS1} + 0.810 \ln \text{LAB}$
$(-2.35) \quad (1.33) \quad (2.26)$
$+ 0.512 \ln \text{PEC} - 0.76 \text{ D80}$
$(2.10) \quad (-2.38)$
$R^2 = 0.99$

(4) $\ln \text{WP1} = 16.111 + 93.907 (\text{MM2/GDP}) + 0.482 \text{ WEP}$
$(5.86) \quad (2.41) \quad (4.08)$
$R^2 = 0.99$

(5) $\ln \text{TX} = -15.355 + 3.082 \ln \text{GDP} - 1.718 \ln \text{RPX}$
$(-3.54) \quad (13.95) \quad (-3.71)$
$R^2 = 0.99$

(6) $\ln \text{TM} = -12.641 + 1.923 \ln \text{GDP} + 0.389 \ln \text{PDO}$
$(-6.42) \quad (9.29) \quad (7.03)$
$R^2 = 0.99$

SIMULATION ANALYSIS

The estimation equations are used for within-sample simulation and the projection of 1986 values to evaluate the impacts of sudden oil price changes. Table 8.2 presents the results of the evaluation of the energy/economy model. Except for total exports, the remaining equations have a root mean square percentage error of less than 15 percent. Primary energy consumption and GDP equations produce better fits.

This section reports the simulation results of the year 1986 under several "no oil shock" scenarios. In order to investigate the impacts of the 1986 oil price drop, scenario 1 assumes no oil price drop in 1986. Scenario 2 assumes no second oil shock (the price of oil increases at an annual rate of 5.2 percent after 1975). Scenario 3 assumes no oil shocks at all; using the actual prices of oil for 1972 and 1986, oil prices between 1972 and 1986 are interpolated. Tables 8.3, 8.4, and 8.5 present the results of the three simulations.

Table 8.2
Errors of the Energy-Economy Models: Within-Sample Simulation, 1971–85

Endogenous variables	MAE	MAPE	RMSE	RMSPE
WEP	7.059	0.1224	9.098	0.1440
PEC	2,056.387	0.0623	2,524.816	0.0786
GDP	113.013	0.0360	1,288.030	0.0438
WPI	3.495	0.0686	4.054	0.0937
TX	925.985	0.1515	1,300.578	0.2004
TM	693.939	0.0760	993.288	0.0919

MAE (mean absolute error) $= \frac{1}{n} \Sigma \, |S_t - A_t|$

MAPE (mean absolute percentage error) $= \frac{1}{n} \left(\frac{\Sigma \, |S_t - A_t|}{A_t} \right)$

RMSE (root mean squared error) $= \sqrt{\frac{1}{n} \Sigma \, (S_t - A_t)^2}$

RMSPE (root mean squared percentage error) $= \sqrt{\frac{1}{n} \frac{\Sigma \, (S_t - A_t)^2}{A_t^2}}$

Where n = number of observations.
A_t = actual value at time t.
S_t = simulated value at time t.

Source: Shin, Jeong-Shik, "An Energy-Economy Model of Korea," Korea Energy Economics Institute, Seoul, Korea, 1987.

Table 8.3
Energy-Economy Interaction Simulation Results with Oil Price Scenario 1

Endogenous variables	1986 Actual value	Actual price (A)	Scenario 1 (B)	(A-B)/A × 100
WEP	117.6	95.6	127.8	+33.7
PEC	61,065.0	66,981.0	65,274.0	-2.5
GDP	61,166.0	61,327.0	60,521.0	-1.3
WPI	125.5	114.0	130.3	+14.3
TX	30,600.0	35,292.0	33,882.0	-1.0
TM	27,841.0	29,872.0	33,628.0	+12.6
Balance of trade	2,759.0	5,420.0	254.0	-95.3

Source: Shin, Jeong-Shik, "An Energy-Economy Model of Korea," Korea Energy Economics Institute, Seoul, Korea, 1987.

Table 8.4
Energy-Economy Interaction Simulation Results with Oil Price Scenario 2 (assuming no second oil shock)

Endogenous variables	1986 actual value	Actual price (A)	Scenario 2 (B)	(A-B)/A × 100
WEP	117.6	95.6	95.5	-0.1
PEC	61,065.0	66,981.0	74,569.0	+11.3
GDP	61,166.0	61,327.0	64,792.0	+5.7
WPI	125.5	114.0	111.2	-2.5
TX	30,600.0	35,292.0	41,802.0	+18.4
TM	27,841.0	29,872.0	33,179.0	+11.1
Balance of trade	2,759.0	5,420.0	8,623.0	+59.1

Source: Shin, Jeong-Shik, "An Energy-Economy Model of Korea," Korea Energy Economics Institute, Seoul, Korea, 1987.

Table 8.5
Energy-Economy Interaction Simulation Results with Oil Price Scenario 3 (assuming no oil shocks)

Endogenous variables	1986 actual value	Actual price (A)	Scenario 3 (B)	(A-B)/A × 100
WEP	117.6	95.6	77.8	-18.6
PEC	61,065.0	66,981.0	88,425.0	+32.0
GDP	61,166.0	61,327.0	70,703.0	+15.3
WPI	125.5	114.0	98.6	-13.5
TX	30,600.0	35,292.0	54,711.0	+55.0
TM	27,841.0	29,872.0	35,449.0	+18.7
Balance of trade	2,759.0	5,420.0	19,263.0	+254.9

Source: Shin, Jeong-Shik, "An Energy-Economy Model of Korea," Korea Energy Economics Institute, Seoul, Korea, 1987.

SIMULATION RESULTS

The simulation results indicate that the Korean economy could have recorded a trade surplus in 1986 even if there had not been a sudden decrease in oil prices. While both GDP and total energy consumption in Korea would have reached much higher levels without the two oil shocks, the rate of increase in energy consumption would have been much higher than that of GDP.

KOREA'S ENERGY FUTURE

The Korean Energy Economics Institute energy model is used to project the evolution of Korea's energy demand through 2001 and to derive the resulting policy implications.

Key Assumptions

The energy forecast presented here is taken from the long-term energy forecast research report ''Energy Future for Korea,'' which has been publicized by the Korean Energy Economics Institute. Economic growth assumptions are from the Korean Economic Planning Board's development plans and the Korean Development Institute's study results of the period 1986–2001. The assumptions on industrial structural changes are made on the basis of the study results of the Korea Institute of Economics and Technology.

The GNP growth rate is predicted to average about 7 percent over the period 1986–2001. The manufacturing sector is assumed to grow at 8.6 percent. Less energy-intensive but high-value-added industries (such as the machinery industry) will continue to grow more rapidly than heavy industries. This trend has been observed in recent statistics and is likely to continue.

The imported crude oil price is assumed to remain at its current level of $18 per barrel (1986 constant dollars) through 1990, and thereafter to increase at 3 percent in real terms.

Energy Demand in the Period 1986–2001

Korea's primary energy consumption will increase at an annual rate of 4.8 percent, from 61.065 million metric tons of oil equivalent in 1986 to 123.382 million in 2001. Table 8.6 shows the energy transition in Korea between 1975 and 2001 by energy source and consuming sector.

Petroleum is projected to increase its share in final energy consumption from its present 50 percent to 53 percent in 1991 and to 59 percent in 2001. (The share of petroleum in primary energy consumption is projected to remain stable at 46–47 percent of the total.) The petroleum consumption share will increase despite efforts to substitute alternative fuels. The industrial sector's reliance on petroleum is projected to decline from 55 percent in 1986 to just under 50 percent in 1991 and to 46 percent in 2001. However, the transport sector, which relies primarily on oil, more than offsets the oil substitution in other sectors. Transport is expected to raise its share in total consumption from 15.4 percent in 1986 to 17.3 percent in 1991 and to 20.8 percent in 2001. Thus, despite major efforts to promote conservation and substitute fuels in electric power generation and industry, Korea will remain highly dependent on imported oil.

Table 8.6
Energy Transition in Korea, 1975–2001 (ktoe)

	1975	1980	1986	1991	2001
Primary Energy Consumption					
Petroleum	15,638	26,830	28,498	38,137	57,917
Anthracite	7,556	9,878	12,842	12,771	7,503
Bituminous	519	3,321	10,092	13,304	33,744
Nuclear	0	869	7,078	12,509	18,422
Liquefied natural gas	0	0	69	2,562	2,600
Others	3,841	3,013	2,458	2,740	3,196
Total	27,554	43,911	61,037	82,023	123,382
Final Energy Consumption					
Petroleum	11,009	19,842	24,315	35,127	55,628
Anthracite	7,047	9,105	11,939	12,067	6,986
Bituminous	519	3,321	6,552	9,431	14,093
Liquefied natural gas	0	0	0	556	2,600
Electricity	1,430	2,815	4,843	7,274	13,509
Others	3,420	2,517	1,480	1,550	1,567
Total	23,425	37,600	50,129	66,005	94,383
Final Energy Consumption by Sector					
Industry	8,883	16,362	21,744	28,452	41,728
Transportation	2,224	4,905	7,700	11,417	19,683
Residential/commercial	10,611	14,019	18,329	22,179	27,272
Public	1,707	2,314	2,355	3,956	5,700
Total	23,425	37,600	50,128	66,004	94,383
Industrial Sector Energy Demand					
Petroleum	7,035	10,738	11,897	14,195	19,109
Coal	830	3,653	6,680	9,531	14,173
Liquefied natural gas	0	0	0	216	529
Electricity	1,018	1,971	3,167	4,510	7,917
Total	8,883	16,362	21,744	28,452	41,728

ktoe = Thousand tons of oil equivalent.

NOTE

1. The GDP elasticity concept here is different from the conventional income elasticity of demand, which is defined as the ratio of percentage change in quantity demanded to the percentage change in income when other things (such as prices) are held constant.

NINE

Energy Sector Performance and Economic Development in Taiwan

TSENG TSENG-TSAI

ENERGY SECTOR PERFORMANCE

Energy planning in the Republic of China began in 1968 with the establishment of the Energy Planning and Development Committee. This committee was initially under the guidance of the Council for International Cooperation and Development of the Executive Yuan, but two years later it was renamed the Energy Policy Committee and transferred to the Ministry of Economic Affairs. In November 1979, the Energy Policy Committee was reorganized and renamed the Energy Committee. It was set up under the Ministry of Economic Affairs and was made responsible for the implementation of energy policy and the enforcement of the Energy Management Law.

The Energy Committee is an interagency committee. Its 20 members are scholars and representatives who come from the Council on Economic Planning and Development, Ministry of Finance, Ministry of Communication, National Science Foundation, Taiwan Provincial Government, Nuclear Energy Committee, China Petroleum Company, Taiwan Power Company, and other energy-related bureaus and institutions in the Ministry of Economic Affairs. The executive secretary serves as the chief of staff of the committee.

The Energy Committee members meet three or four times a year. At the meetings, discussions on energy-related programs, such as energy policy, energy investment planning, energy technology development, and the energy conservation program, are held. Energy-related decision making is integrated with economic planning because the members are from the ministries involved in economic planning. Energy investments are reviewed by the Committee on Energy Planning and Policy after being passed at the Energy Committee meeting.

In the Republic of China, the average CIF price of imported crude oil was U.S.$28.45 per barrel in 1985. However, the international oil price slump at the beginning of 1986 has reduced the average import cost to U.S.$15.30 per barrel, a decrease of 46.2 percent. To respond to public opinion and reflect the cost reduction of imported crude oil, the government decreased the domestic prices of petroleum products four times in February, March, and April 1986. The accumulated rates of price decreases were 27, 21, 19, 15, and 10 percent for gasoline, diesel, fuel oil, LPG, and LNG, respectively. Despite continued price decreases in the international oil market after April 1986, the government did not further lower domestic oil prices because of the uncertainty of the movements of international oil prices and Taiwan's lower petroleum product prices relative to its neighboring trade competitors—Japan, Korea, Singapore, and Hong Kong. Moreover, the existing policies of energy conservation and industrial structural adjustment are very important considerations. However, this decision has been generally opposed by public opinion. The continuous pressure for price cuts did not cease until the international oil price started to increase in September 1986.

During this period, no significant change has been made in long-term energy policy. The only change is the increase in customs duty on crude oil from 2.5 percent to 5 percent. The sales tax on domestic oil products has not increased. The government-owned Chinese Petroleum Corporation earned a NT$46 billion profit in 1986.

Because the excess profits of the corporation are estimated to increase significantly in coming years, the government is considering an increase in the sales tax on oil products.

Sources of energy other than oil will continue to increase in the 1990s despite lower oil prices:

- Use of natural gas will progressively increase and will be the major force in interfuel competition because the decline in the price of oil has been accompanied by price reductions in competing fuels, particularly natural gas and coal.
- Nuclear power, despite environmental pressures and a curtailed program in the United States, continues to be developed.
- It is believed that energy conservation investment will continue, because the country does not have any significant indigenous energy resources.

Table 9.1 presents the latest projections of the energy supply and demand balance for 1985–2000 and presents a breakdown for oil, natural gas, coal, hydro, nuclear, and other energy sources.

Despite weak growth in recent years, noncommercial and renewable energy are projected to increase stably over the remainder of the period. This is because of the following:

Table 9.1
Energy Supply and Demand, 1985–2000 (1,000 kloe)

	Energy supply			Energy demand						
Period	Total	Indigenous	Imported	Total	Coal	Petroleum	Natural gas	Hydro	Nuclear power	Others
1985	38,514	4,095	34,419	38,514	8,653	19,671	1,298	1,657	6,885	350
1986	41,702	4,152	37,550	41,702	10,011	22,198	1,141	1,723	6,275	354
1987	43,727	4,125	39,602	43,727	10,832	22,991	1,299	1,511	6,737	357
1988	45,700	4,039	41,661	45,700	10,953	24,719	1,336	1,510	6,819	363
1989	48,374	4,158	44,216	48,374	11,667	26,330	1,533	1,617	6,853	374
1990	50,089	3,288	46,801	50,089	12,427	26,253	2,411	1,689	6,863	446
1991	52,435	3,339	49,096	52,435	13,994	26,019	2,579	1,689	7,655	499
1992	54,579	3,327	51,252	54,579	14,710	27,335	2,713	1,821	7,441	559
1993	56,598	3,385	53,213	56,598	15,630	28,031	2,835	1,942	7,537	623
1994	59,643	3,591	56,052	59,643	17,451	28,999	2,958	1,960	7,575	700
1995	63,188	3,832	59,356	63,188	19,833	29,749	3,100	2,090	7,640	776
1996	65,598	3,987	61,611	65,598	19,416	29,758	3,136	2,276	10,155	857
1997	67,935	4,088	63,847	67,935	20,174	30,922	3,165	2,269	10,399	1,006
1998	71,094	4,270	66,824	71,094	22,420	31,604	3,188	2,228	10,407	1,247
1999	73,726	4,774	68,952	73,726	23,179	32,679	3,229	2,492	10,656	1,491
2000	76,749	5,334	71,415	76,749	24,249	33,461	3,284	2,627	11,325	1,803
Structure of Energy Supply and Demand, 1986–2000 (%)										
1986	100	10	90	100	24	53	3	4	15	1
1990	100	7	93	100	25	52	5	3	14	1
1995	100	6	94	100	31	47	5	3	12	1
2000	100	7	93	100	32	44	4	3	15	2
Annual Growth Rate of Energy Supply and Demand, 1986–2000 (%)										
1986–1990	4.7	-5.7	5.7	4.7	5.6	4.3	20.6	-0.5	2.3	5.9
1990–2000	4.4	5.0	4.3	4.4	6.9	2.5	3.1	4.5	5.1	15.0
1986–2000	4.5	1.8	4.7	4.5	6.5	3.0	7.8	3.1	4.3	12.3

kloe = Kiloliters of oil equivalent.

Source: Energy Committee, Ministry of Economic Affairs, Republic of China, Taipei, 1987.

- Growing requirements for solar heating
- Increasing biomass requirements
- Expected increases in oil prices, beginning in the mid–1990s.

EVIDENCE OF STRUCTURAL CHANGES IN ECONOMY AND ENERGY/ECONOMY RELATIONSHIPS

Agriculture, Industry, and Services

In 1970, agricultural production accounted for only 16.7 percent of real GDP; industrial production (including mining, manufacturing, construction, and public utilities) and service production accounted for 39.2 percent and 44.1 percent, respectively. In 1986, agriculture's share further declined to 5.6 percent, industry increased to 51.6 percent, and services accounted for 42.8 percent. During the same period, real GDP in agriculture, industry, and services grew at annual rates of 1.3, 10.4, and 8.3 percent, respectively.

Structure of GDP by Manufacturing Sectors

During the brief interlude between the 1973–1974 and 1979–1980 oil crises, the structure of GDP by manufacturing sectors changed significantly in the Republic of China. In 1970, the 7 largest shares among 14 manufacturing sectors were food (24.7 percent), petroleum and coal products (19.6 percent), textiles (11 percent), chemical products (9.4 percent), wood products (5.9 percent), nonmetallic minerals (5.8 percent), and electronic machinery (5.7 percent). In 1985, the big seven were textiles (15.7 percent), chemical products (14.8 percent), electronic machinery (12.3 percent), processed food (11.8 percent), petroleum and coal products (6.7 percent), basic metals (6 percent), and transportation equipment (5.4 percent). It is quite surprising that energy-intensive sectors such as chemical products, basic metals, and textiles grew so rapidly during the 1970s and mid–1980s. Between 1970 and 1985, the annual growth rates of these sectors were 15.4, 19.3, and 14.7 percent, respectively. In the next section we will discuss government policies that induced such structural changes.

During the two oil crises, the energy intensity per unit of real GDP increased by 10 percent in the Republic of China. This was partially due to long-term planning to develop the petrochemical and heavy industrial sectors, including the establishment of an integrated steel mill to upgrade the industrial structure. Unfortunately, these industries were quite energy intensive as well as fast growing. Between 1973 and 1979, the growth rate for 18 petrochemical products averaged 27 percent per year, while the output of the steel industry grew by 15 percent per year.

Why were these energy-intensive sectors being developed at a time when the government actively sought to restrict aggregate energy demand growth in the economy? To understand the economic rationale behind this step, we must take a very brief look back at the industrial structure of the 1950s and 1960s. Even though high growth during these two decades was accompanied by impressive industrial structural change, for the most part, simple processing and export operations fueled by imports of basic materials and intermediates characterized the economic structure. Moreover, the shortage of imported basic materials was one of the most important factors underlying the inflation that arose after the first oil crisis. To eliminate the problem of inflation-feeding supply shortages, the government undertook the development of indigenous supplies of basic materials and intermediates. The strategy, calling for the development of the energy-intensive heavy industry and petrochemical sectors, was most vigorously pursued in the five-year span between the two oil crises.

Energy Price Policy

1973–1979. When the first oil crisis hit, the Chinese Petroleum Corporation did not reflect its true costs in the prices of its petroleum products in order to follow the government's strategy of facilitating continued prosperity. In response to the 216 percent increase in the cost of imported crude oil in 1973–1974, the domestic price index for petroleum products increased by 76 percent, which was lower than the price increases for other major fuels during this period—the prices of coal, natural gas, and electricity were raised by 208, 245, and 78 percent, respectively. During 1974–1979, the average annual increase in imported crude oil prices was 11.4 percent. The yearly growth rates of domestic energy prices during the same period were petroleum products, 3.4 percent; natural gas, 11.2 percent; and electricity, 6.6 percent.

1979–1986. After the persistent and severe global recession brought on by the second oil crisis of 1979, the government had to increase the price of energy to reflect the true costs and to implement an energy conservation policy. The price of crude oil imports grew at an annual rate of 42.3 percent during the second oil crisis (1979–1981). During the same period, the domestic prices of petroleum products were adjusted upward by a sizable 40 percent per year. Other major fuels were also given large annual price increases—coal, 31 percent; natural gas, 50 percent; and electricity, 39 percent.

Macroeconomic Policy

1974–1979. The government's initial response to the sudden and swift upward movement in the price of imported oil in 1973 was to try to regulate prices and continue its subsidy program; both these measures, however, were found to spur fear of future inflation among the public, thus adding to inflation, and were soon abandoned.

The Economic Stabilization Program was declared in January 1974. The highlights of this program were the following:

- A substantial increase in interest rates and a corresponding tightening of credit
- An upward adjustment of prices for petroleum products, electricity, transportation services, wine, and tobacco
- A reduction in government subsidies for imported soybeans and wheat
- The partial liberalization of price controls through the replacement of price ceilings with negotiated prices.

The program's effect was almost immediate. Prices jumped in February but soon stabilized on a higher plateau. Price negotiations were eliminated in June. In sum, this once-and-for-all price adjustment fought successful battles on two fronts: (1) it liquidated the public's psychology of inflation by allowing prices to rise almost overnight to high levels, where they then remained; (2) contractive monetary policy checked the growth of demand and double-digit inflation was stopped.

Though inflation was stopped, the economic growth rate dropped from 12.8 percent in 1973 to 5.95 percent and 2.13 percent in the first two quarters of 1974. The government poured funds from its coffers into the "Ten Great Projects" to offset declining investment by the private sector. As a result, the rate of increase in 1974 for investment in fixed capital formation was 9.2 percent, about equal to the 1973 level. In addition, the following measures were taken to push the economy out of its recession:

- Lower import tax rates
- Reduced interest rates and bank reserve requirements
- Refinance of export loans
- Resumption of import tax refunds.

1979–1986. In view of the failure of price-control policy in 1973, the government adjusted the prices of domestic oil products and electricity soon after the price of imported oil increased during the second oil crisis. Meanwhile, macroeconomic policies were deliberately used to deal with domestic inflation caused by the huge increase in oil prices. The policies adopted were cutting taxes, tightening the money supply, and cutting public expenditures, especially capital formation by public enterprises.

The second oil shock struck a heavy blow to the domestic economy. As a result, in the Four-Year Plan covering 1982–1986 (released in 1981), some fundamental policy changes were evident:

- The expansion of energy-intensive industries (such as aluminum, zinc, and lead) was to be restrained, and incentives were offered to encourage the development of these industries abroad as part of joint-venture projects.
- The expansion of steel and copper refining industries was to be carried out only at the same pace as the development of the electrical and nonelectrical machinery industries.

- The development of the petrochemical industry was to continue, but at a pace adequate to supply only the needs of the domestic market and the material requirements of the export sector.
- Increasing output from low-energy, high-value-added strategic industries (the machinery manufacturing and electronics sectors) was a key priority to be pursued vigorously in the 1980s.

These policies help to illuminate the more general approach to energy conservation now held by economic planners. The new orientation has evolved from past economic management experience and from the conditions of resource scarcity in the domestic economy and the ever present threat of a third oil crisis.

The government is now interested in restricting oil consumption in the following ways:

- Developing a sophisticated energy-saving program that promotes energy conservation through the use of energy-efficient machinery and equipment
- Improving the industrial structure by expanding the overall output shares of energy-efficient, skill-intensive, and high-value-added industries
- Diversifying energy sources for domestic consumption, particularly in the area of electric power generation.

The Four-Year Plan for 1982–1986 stipulates a target of increased efficiency in energy use of 10.5 percent in the manufacturing sector.

The changes in energy intensity of the economy, which is indicated by the ratio of energy use to GDP, were attributed partially to the changes in the composition of GDP. In general, the larger the share of energy-intensive industries, the greater the energy intensity in the seven main sectors will be. During 1970–1986, manufacturing climbed from a 31 percent share to a 41 percent share, and transportation rose from 3.7 percent to 6.1 percent. This moves the energy intensity of the economy upward.

The energy intensity of the economy was also affected by the government's industrial policy and energy demand management policy. Although the government's campaign to set up the petrochemical industry and the giant steel mill during 1974–1978 was reasonable, it drove up the energy intensity of the economy to some extent. Finally, the changes in energy prices with respect to prices of capital, labor, and nonenergy intermediate inputs were, of course, a very important factor in explaining the changes in energy intensity. Owing to the increased GDP share of the manufacturing and transportation sectors, which consume more energy than the other sectors, and the government's industrial policy during 1974–1978, energy productivity in terms of 1981 New Taiwan dollars per liter of oil equivalent decreased from 71 in 1970 to 65 in 1980, despite the fivefold increase in the domestic petroleum price index.

However, the real-cost energy pricing policy and the other energy demand management policies taken by the government in response to the second oil crisis did effectively check the energy intensity of the economy from further growth. This was shown by its drop in the period from 1979 to 1982. After the government markedly adjusted the prices of petroleum products, natural gas, and electricity upward in 1979–1981, energy productivity increased from 64 in 1979 to 71 in 1982.

In addition, in August 1980 the government moved to fight energy waste with the passage of the Energy Management Law, which provides the legal basis for energy-saving policies.

Examining the energy intensity at the sectoral level shows that the three factors mentioned are also factors that contributed to changes in energy intensities.

The significant government investments in the energy sector during 1973–1986 were the following:

- Six-unit nuclear power plant (total capacity, 5,144 MW)
- Four-unit oil-fired power plant (total capacity, 2,000 MW)
- Four-unit coal-fired power plant (total capacity, 2,100 MW)
- Four-unit pumped-storage power plant (total capacity, 1,000 MW)
- Taoyuan refinery (130,000 barrels per day)
- No. 3 and No. 4 naphtha cracking unit
- Offshore natural gas exploitation project
- LNG importing project.

CHANGING OIL PRICES AND ECONOMIC DEVELOPMENT

An energy pricing policy that fully reflects real costs is an effective instrument for promoting energy conservation. But this is only true in the case of increasing real costs. In the case of decreasing real costs, such as the decreasing price of imported crude oil, full-cost energy pricing is no longer an effective policy tool for promoting energy conservation. Instead, it encourages the consumption of energy and, hence, lowers the energy efficiency of the country. The price of imported crude oil has tailed off from U.S.$33.90 per barrel in 1981 to U.S.$28.45 per barrel in 1985 and U.S.$15.30 per barrel in 1986. The decreasing trend in crude oil prices seems likely to last for several years, and causes new conflicts concerning the goals of energy pricing policy. An immediate and full reduction in domestic energy prices in response to the fall in imported oil prices may discourage energy conservation. But maintaining unrealistically high energy prices would certainly cause the price competitiveness of exports to weaken and would retard economic growth. A trade-off between the two extremes to simultaneously achieve energy conservation and economic growth is difficult to make.

From 1981 to 1985, the price of imported crude oil declined by 5.8 percent per year. In response, the domestic prices of petroleum products were reduced by 2.3 percent and the electricity rate by 1.3 percent per year. This was a very mild price adjustment. From 1985 to 1986, the price of imported crude oil declined by 46.2 percent. In response, the domestic prices of petroleum products were reduced by 22.4 percent and the electricity rate by 7.7 percent. The reasons provided by the government for such a mild adjustment in domestic energy prices were the following:

- To continue the active promotion of energy conservation
- To provide funds for the "Fourteen Key Projects" through the increase in profits of the Chinese Petroleum Corporation and the Taiwan Power Company.

In addition, the government claimed that the overall energy prices in Taiwan were not higher than those in Japan, Korea, Hong Kong, or even Singapore. This seems to reconfirm that since the outbreak of the second oil crisis, energy conservation has become one of the most important considerations of the government in energy pricing policy.

In a report titled *Petroleum Price Forecasting System*, published in March 1987, SRI International claimed that worldwide prices of crude oil would stabilize for a few years but would rise again at the end of the 1980s. If this forecast is right, maintaining the present energy conservation policy is a wise decision. Nevertheless, it is suggested that in a highly open economy like Taiwan's, the prices of energy should not be kept on a higher level than those of its export competitors. Otherwise, the price competitiveness of Taiwan's exports will be weakened and its economic growth will be impeded.

CURRENT ECONOMIC DEVELOPMENT PLANNING AND STRATEGIES

In view of the current problems, the strategy for economic development in the long term is to make strong efforts toward economic liberalization and internationalization.

For the short term, Taiwan is actively engaged in improving the investment environment to stimulate private investment in order to accelerate industrial improvement, increase economic productivity, diversify export markets under a good trade image, and continue to promote balanced economic and social development. Furthermore, after the completion of the "Ten Major Projects" and the "Twelve Minor Projects," the government decided to carry out the "Fourteen Key Projects," beginning July 1986.

While continuously improving economic construction in the future, the supreme principle of economic development is still to follow Dr. Sun Yat-sen's Principle of the People's Livelihood by raising living standards, taking care of the people, and letting everybody share in the fruits of economic development.

Having one of the world's smaller, less self-reliant economies, Taiwan has long recognized the importance of strong and stable trade and investment ties with the rest of the world. Accordingly, its future economic and energy strategies will be continuously placed in a global framework. It is the belief of the Taiwan government that all countries, irrespective of differences in oil production, consumption, and level of development, basically find themselves in the same situation of international interdependence. And there is everywhere an acute awareness of the necessity to achieve stability of energy production and demand as well as security of supply—in peace and in times of crisis.

Consequently, Taiwan always seeks plentiful and correct information regarding developments that influence the supply and price of resources that involve technological innovations in production, consumption, or environmental protection. If the common interest is to draw economic and energy plans rationally and realistically, there is no better choice than to participate actively in international dialogues on economic and energy problems.

The Republic of China has had much experience working in collaboration with other countries and international organizations, and has enjoyed fruitful results in building the foundations for reliable energy forecasts and creating the basis for rational evaluation of trends in energy production, consumption, stockpiling, technology, and security of supply. It will strengthen this practice in the future. At the same time, while acting in accordance with its national interests, the Republic of China recognizes its international responsibility and offers its services in the interests of worldwide economic and energy development.

In summary, despite its heavy dependence on imported energy during the decade of "energy crises," the Republic of China performed quite well in the following aspects:

- Impressive economic growth rates—8.5 percent per year real GDP growth during 1970–1986
- Relatively low increase in domestic prices in general—6.5 percent per year in terms of wholesale prices during the same period
- Absence of external indebtedness for oil import financing—in fact, foreign exchange reserves increased from U.S.$1.55 billion in 1974 to U.S.$60 billion in June 1987.

It is believed that the government policies mentioned here could be adopted by other energy-importing countries.

PART IV
Country Case Studies: Southeast Asia

TEN

Energy Plans and Strategies: Malaysia

SAHATHAVAN MEYANATHAN

INTRODUCTION

This chapter briefly examines Malaysia's energy plans. In particular, it focuses on the energy sector plans, macro energy sector linkages, policy responses to recent developments in oil markets, the institutional setup in the energy sector, and the cooperative endeavors in the field of energy between Malaysia and the other countries of the region.

REVIEW OF ENERGY SECTOR PLANS: THEIR KEY OBJECTIVES AND ASSUMPTIONS

Energy planning in a very loose sense started with the enunciation of the five-year development plans that have been in force since independence in 1957. The allocation in the plans for the energy sector constituted elements of energy planning. A sum of M$993 million was allocated for the energy sector in the Fourth Malaysia Plan, for 1981–1985, a thirteenfold increase from the previous plan. Most of the allocation was for hydro power, rural electrification, and transmission. For the Fifth Malaysia Plan, the allocation is M$6.385 billion over the years 1986–1990 for hydro power, thermal, and gas electricity generation, rural electrification, and transmission and distribution.

Although an overall energy policy was formulated in 1979, it constituted no more than broad guidelines on long-term objectives and strategies, and these were carried into the Fourth Malaysia Plan. The major thrust of the policy was energy diversification based on the four-fuel strategy, "namely oil, hydro, gas, and coal and aimed at ensuring reliability and security of supplies while reducing the dependence on oil in energy consumption." The objectives were the following:

- Supply—" . . . to provide the nation with adequate and secure energy supplies by reducing our dependence on oil and by developing and utilizing alternative sources of energy" (Government of Malaysia 1982, p. 12). "A four-fuel energy strategy based on oil, hydro, gas, and coal has been formulated towards diversifying our energy base as well as guaranteeing assured energy supplies for our continued growth."[1]
- Utilization—Measures aimed at efficient utilization of energy.
- Environment—The achievement of the above two objectives was not to be at the expense of the environment.

Around the same time, specific policies were formulated for the oil sector. They included the following:

- Making available adequate supplies at reasonable prices to meet domestic consumption
- Opening up new opportunities for the establishment of heavy industries
- Increasing revenue and export earnings by increasing exports of oil and gas
- Malaysian representation in terms of ownership, management, and control in petroleum exploration, marketing, and distribution
- Optimum exploitation of oil and gas in relation to the need for conservation and protection of the environment.

Further, the National Depletion Policy, introduced on June 1, 1980, restricted the production of crude oil to 1.75 percent of oil initially in place for major fields.

Energy conservation is mainly directed at oil. Electricity tariff rates were structured to reduce waste, and the government also has imposed a higher road tax for cars with engines exceeding 1500cc. Fiscal measures such as an accelerated depreciation allowance on expenditures on plant and machinery for purposes of modernization were also taken. Effective in 1982, firms were permitted to generate power for their own use from industrial wastes.

In November 1983, the National Energy Planning Study was initiated, with the objectives of developing a comprehensive energy planning approach. This included the development of tools and data bases for planning, policy formulation, and decision making. The aim was to institutionalize energy planning as a continuous exercise. In addition, the Power Sector Investment Study, undertaken in 1985, examines the magnitude of national power sector investments over a time frame up to the year 2010.

In the latest official plan (Fifth Malaysia Plan), published in March 1986, the objectives of the Fourth Plan vis-à-vis the energy sector will continue to be pursued. The concept of integrated energy planning is emphasized, with a view to ensuring systematic and optimal development of the sector. The only targets provided in the plan relate to the following:

- Supply of primary energy
- Electricity generation, by source

- Energy demand, by source
- Energy demand, by sector, for the period 1986–1990.

The average annual growth rate of the supply of primary energy is targeted at 6.8 percent over the period 1986–1990. In terms of share in the total primary energy supply, crude oil's contribution is expected to decline, falling to 41.7 percent in 1990 (compared with 53 percent in 1985). The dependence on natural gas, 18.1 percent in 1985, is targeted to increase to 20.8 percent in 1990, implying an average annual growth rate of 7.9 percent. The share of petroleum products in the primary energy supply is targeted to grow by 10.9 percent over the period 1986–1990. Coal will increase its share from 2.8 percent in 1985 to 8.7 percent in 1990. By 1990, the share of gas in electricity generation is targeted to reach 50.6 percent; coal, 15.5 percent; and hydropower, 19.7 percent; oil's share is to be reduced to 14.2 percent. The target for overall growth of energy demand for the period 1986–1990 is 7 percent per year. In terms of energy mix in total demand, the contribution of coal is targeted to grow at 16.6 percent over the same period, followed by electricity, growing at 8.3 percent per year; the average annual growth rate of petroleum products is targeted at 6.3 percent. The share of gas was 2.4 percent in 1985, and is projected to grow at 9.7 percent annually, accounting for 2.8 percent of the total energy demand in 1990.

During the Fifth Malaysia Plan, the thrust for the petroleum industry will be toward more intensive prospecting and exploration activities to augment known resources that are being depleted. Positive measures have been undertaken to promote more intensive exploration activities, which have slackened in the face of falling crude oil prices and global recession. Exploration drilling from the mid–1960s to the mid–1980s has indicated that future discoveries are likely to be found in smaller deposits. Thus, terms and conditions of production-sharing contracts have been revised to encourage private sector investment. Foreign participation is promoted, since local expertise is still not well developed.

A sum of M$1.008 billion is allocated for Petronas under the Fifth Malaysia Plan, of which M$1 billion will be spent by Petronas Carigali, the exploration and production arm of Petronas, in the development of the Dulang field, the Duyong phase 2 project, and exploration drilling. The Dulang field is expected to come onstream by 1998.

The key factors in defining energy objectives and targets in Malaysia have been the following:

- Domestic energy resource endowment
- Target GDP levels
- The share of industry in GDP

CURRENT THINKING ON CRITICAL PLANNING INPUTS

The current view of the policymakers on the future trend of global oil prices is generally optimistic over the longer term. There is a consensus that oil prices will recover after 1990. Over the medium term, during the period 1987–1990, oil prices are expected to remain low, in the region of U.S.$16–19, based on the assumptions in the Fifth Malaysia Plan. Projections of oil prices for the period 1990–1995 are in the range of U.S.$20–25. Prices beyond 1995 are not expected to fall below the level of U.S.$20–25. The 1985 Bank Negara Annual Report (p. 47) has estimated an average price of U.S.$20 for crude oil until the end of 1986.

A survey of the representative views of policymakers regarding the growth rate of the domestic economy indicates an annual rate of 1 to 3 percent for the period 1986–1987, comparable with the growth rate of 2.8 percent recorded for 1985, the lowest since 1976. A growth rate of 3.3 percent for 1985–1986 was estimated by Bank Negara (Bank Negara Annual Report, 1985). The Fifth Malaysia Plan projected a 5 percent growth rate for the economy over the period 1986–1990.

The percentage share of industry in the GDP in 1985 was 19.1 percent. The Fifth Malaysia Plan projected a conservative increase in the percentage share of industry to 20 percent over the period 1986–1990, while the Industrial Master Plan estimates is 25 percent for the period 1990–1995.

With the exception of the 1986–1987 forecast of global oil prices, which we believe will average about U.S.$16–19, we believe that the ranges are realistic even if the specific point estimates may not be.

POLICY RESPONSES TO RECENT DEVELOPMENTS IN INTERNATIONAL OIL MARKETS

Macrolevel Adjustment

The recent fall in oil prices has brought about no direct or immediate reaction by the government. The fiscal, monetary, and expenditure/investment policies have been undergoing adjustment since mid–1982 to address the deficits in the budget and current account. These problems were addressed through a medium-term adjustment program aimed at bringing spending more in line with income and paying more attention to the mounting external debt. The measures essentially relied on the reduction and rephasing of public expenditure over the short term and privatization over the medium term. The change in government expenditure was from a positive 27.8 percent in 1981 to 4.1 percent in 1982, and to negative 0.4 percent and 0.2 percent in 1983 and 1984, respectively. In 1985, it fell by 1.4 percent. The overall government deficit as a percentage of GNP, which stood at about 20 percent in 1982, was 13 percent in 1983, 10.1 percent in 1984, and 6.8 percent in 1985. Since 1980, crude petroleum has been the nation's leading foreign exchange earner. Export earnings of crude petroleum increased from M$6.709 billion in 1980 to M$8.727 billion in 1984 and to M$8.9695 billion in 1985, due entirely to an increase in export volume. Crude petroleum's current share of total gross export earnings stands at 23.5 percent.

Fluctuations in oil prices are bound to affect the Malaysian economy, given its substantial contribution to GDP and oil-related revenue. The strategy to boost oil production by 18.6 percent, from 430,000 barrels per day (b/d) in 1985 to 510,000 b/d in 1986, is aimed at sustaining oil revenues needed to finance growth and operating expenditures. The Treasury's forecast of an 18 percent production volume expansion to bring a 16 percent increase in export earnings is generally regarded as overly optimistic. At the prevailing price of U.S.$13.50 per barrel, and assuming the targeted production level of 510,000 b/d is achieved, the earnings from petroleum are expected to amount to only M$4.9 billion, compared with the earlier forecast of M$9.5 billion based on U.S.$26 per barrel, resulting in a shortfall of M$4.6 billion in oil export earnings.

Oil-related revenue comprising royalties, export duties, and income taxes on petroleum companies consistently accounted for around 22 percent of total federal revenue for the period 1980–1985. Generally, the increase in total oil revenue has not been proportional to the expansion in production volume. An increase in production of 16 percent in 1984 brought in only a 7.5 percent increase in revenue, despite the relatively stable prices averaging U.S.$30 per barrel.

Apart from revenue contributions, the oil industry's inflow of oil-related investments constitutes a major component of total private investment. Since 1983, oil industry investments have been contributing around M$2 billion a year, about 20 percent of total private investment. A sharp decrease in oil industry investments would have adverse effects on total private investment.

Oil Production Level Adjustment

At the beginning of 1986, the government set the production target at 510,000 b/d. The production was raised by almost 19 percent over the 1985 level (446,100 b/d) to help cushion the fall in export earnings and government revenue.

In supporting OPEC's moves to stabilize the oil market, the government announced on May 20, 1986, a planned 20 percent production cutback to 459,000 b/d should OPEC members agree to reduce their total output to 16 million b/d, from the present 17 million b/d, in order to prop up falling oil prices (*New Straits Times*, May 23, 1986). The proposed cutback may cause the government to adjust the development budget as a result of lower revenue. The government has anticipated a further M$300 million loss in earnings from such a cutback in crude oil production.

Falling oil prices, competition from neighboring countries in contract terms, and smaller discoveries prompted revisions in the production-sharing contracts. The new terms for production-sharing contracts were approved by the Malaysian Cabinet on December 4, 1985, and were announced the following day. The relaxed terms for oil exploration will be on a sliding scale based on average daily production. Under the new terms, contractors of small fields can claim up to 50 percent of gross production of crude oil, which allows for more speedy cost recovery. The oil profit split[2] for the first 10,000 b/d of production will be 50:50 between Petronas and the contractor. The ratio for the next 10,000 b/d is 60:40 in favor of Petronas, and all production in excess of 20,000 b/d will be split 70:30 in Petronas' favor. The profit formula extends only to cumulative production from the contract area. All production exceeding this limit will be subject to the oil profit split of 70:30, favoring Petronas (*Asian Oil and Gas*, January 1986).

As an additional incentive, all bonus payments have been waived. The elimination of signature, discovery, and production bonuses is expected to spur exploration activities. Prior to the announcement of the new terms, contractors had to pay a signature bonus plus a discovery bonus of M$2.5 million, and production bonuses of M$5 million per 50,000 b/d. Terms for the development of gas discoveries also are substantially relaxed. The percentage allowed for cost recovery is 60 percent, while all profit gas will be split 50:50 for the first 2 trillion cubic feet of production from the contract area. Once this level is reached, the 65:35 ratio in favor of Petronas will be resumed. Under the new terms, there is also a requirement that Petronas Carigali have a minimum of 15 percent equity in any contract area awarded.

Oil Export Price Adjustment

The collapse in crude oil prices forced Petronas to revise its crude prices as prices on world spot markets plummeted. Malaysian crude prices have been set with reference to OPEC prices and spot prices at Rotterdam. Petronas announced an across-the-board price cut of U.S.$4, effective February 1, 1986. As a result, the prices of the various grades were as follows:

Tapis	U.S.$23.90
Miri light	U.S.$23.25
Labuan crude	U.S.$23.60
Bintulu Crude	U.S.$22.55
Tembungo crude	U.S.$23.90

In March 1986, Malaysia further lowered the price of Tapis blend, its highest-volume grade, by U.S.$6.80 a barrel, to U.S.$17.10 a barrel. Following persistent soft market conditions, Petronas made three price cuts within three months, bringing the main grade Tapis to an all-time low of U.S.$13.50 in May 1986 (*New Straits Times,* February 23 and May 23, 1986). Although about 80 percent of Malaysian crude is sold on a long-term basis, over the longer run, spot prices will affect contract prices.

Retail Price Adjustments

On the domestic side, retail prices of petroleum products, which started to fall in August 1985, in response to weakening crude prices, continued to fall in 1986. From March 1986, the retail price of premium gasoline was reduced by 12 cents a liter, to 99 cents, for Peninsular Malaysia. The new prices of regular gasoline and LPG were 11 cents lower. The prices of diesel and kerosene dropped 9.6 cents and 4.8 cents a liter, respectively. On April 1, retail prices dropped further by 4 cents and 5 cents for premium and regular gasoline, respectively (*Business Times*, April 1, 1986). The domestic price of LPG was lowered by 3 cents, while diesel and kerosene prices were reduced by 3.1 and 4.8 cents, respectively.

A further reduction was announced on June 1, 1986. LPG prices decreased by 4 cents per kilo. The price of regular gasoline fell by 1 cent, while kerosene prices decreased by 1.6 cents per liter. The prices for premium and diesel remained unchanged. With this change, the retail price of regular gasoline was 86 cents; and kerosene, 52.5 cents per liter in Peninsular Malaysia (*New Straits Times*, May 31, 1986). Thus, compared with November 1985, prices had fallen by 15.3 percent for premium gasoline, 16.5 percent for regular gasoline, 21.7 percent for diesel, 17.5 percent for kerosene, and 13.6 percent for LPG. It ought to be noted that, with the fall in crude prices, duties on petroleum products had risen to generate additional revenue for the government. Since October 1, 1984, the duty on premium and regular gasoline increased by 6.02 cents a liter to the current 44.52 cents a liter. In the case of diesel, the import duty, which stood at 0.73 cents a liter over the period October 1984 to February 1986, shot up to 8.53 cents in May and dropped to 7.89 cents effective June 1986. The duty on LPG was 4.46 cents per kilo between October 1984 and April 1986. In May 1986, the duty was raised to 6.46 cents and was increased even more substantially to 15.93 cents per kilo in June 1986. As of May 1986, the import duty on kerosene was 7.4 cents a liter, up 5.9 cents from the previous value of 1.5 cents a liter between October 1984 and April 1986. Prior to October 1984, the duty on kerosene was 0.22 cents per liter.

Energy pricing at the retail level in Malaysia can be considered as a policy variable. The changes in the price of crude, which is used as throughput in Malaysian refineries, have been reflected in retail prices of petroleum products. However, the licensing of gasoline, diesel, and kerosene, and their inclusion in the list of controlled items as of 1974, caused difficulties in adjusting prices to market conditions. The higher import costs during this period led to the introduction of subsidies to kerosene in 1976 and to diesel in 1977. These subsidies were partly recovered from taxes imposed on regular and premium gasoline. As the cost of the subsidies rose, on January 1, 1984, the government completely withdrew them. Subsidies are now provided only to LPG in the states of Sabah and Sarawak, at a rate of 33.22 cents per kilo. Prior to April 1983, the oil companies' proposals for changes in retail prices had to be submitted to the

Ministry of Trade and Industry for approval. Since this proved cumbersome and time-consuming, the government, as of April 2, 1983, instituted a new formula that relates retail prices to the following:

- Product cost, including exchange rate changes
- Margin of refiners
- Dealers' commission
- Duty

This is computed on a monthly basis; should there be a change, it is announced on the first of the month.

Electricity Tariffs Adjustment

The decline in the cost of electricity generation resulting from lower fuel cost and increased efficiency led the National Electricity Board to announce reductions in its tariff rates for all commercial, industrial, and mining consumers, effective September 1, 1985. The reduction of tariffs ranged from 4 to 12 percent for medium- and high-voltage commercial, industrial, and mining sectors, and from 50 to 53 percent for off-peak tariff rates for these categories of consumers. The new rates are aimed at stimulating private sector investment (Fifth Malaysian Plan, 1986). Increased discounts were previously provided to large industrial consumers. In 1980, this was directed at heavy industries; in 1984, at the textiles sector.

From March 13, 1986, a further reduction in electricity rates for industrial, mining, and hotel consumers in the form of a 10 percent discount on their monthly bills took effect. Within the category of industrial consumers, special consideration was given to approved textile industries and the Perwaga steel mill, which will enjoy an overall 20 percent discount on their monthly bills.

Change in Energy Plan

Although there has been no change in the objectives of the latest energy plan as a result of oil price developments, the momentum generated thus far toward oil substitution may slacken in view of lower oil prices and lower development funds. More specifically, the allocations made in the Fifth Malaysia Plan for the energy sector may be adjusted downward, implying the phasing down of the larger power stations, including the Bakun Project.[3] Despite the weak oil market, Petronas is expected to drill some 120 exploration wells over the period 1986–1990.

INSTITUTIONAL SETUP IN THE ENERGY SECTOR

At present there is no separate department or agency dealing solely with energy matters. Decision making on energy matters is widely dispersed within the government. Matters pertaining to power generation lie within the province of the Ministry of Energy, Telecommunications, and Posts; pricing of petroleum products is determined by the Ministry of Trade and Industry; the development and promotion of the petroleum industry is governed by Petronas and the Petroleum Development Unit of the Prime Minister's Department; coal exploration and exploitation activities are determined by the Geological Survey Department and the Mines Department under policy directives of the Ministry of Primary Industries; research and promotion of nuclear and solar energy come under the purview of the Ministry of Science, Technology, and Environment.

Coordination in energy planning is achieved through the Cabinet Committee on Energy, chaired by the prime minister. This committee is assisted by the Implementation and Coordination Committee, with members drawn from various energy-related ministries and agencies, and is concerned with supply and its sources over the short, medium, and long terms, as well as evaluation of existing policies with the aim of formulating a national energy policy. The secretariat of the Coordination and Implementation Committee is the Energy Unit of the Ministry of Energy, Telecommunications, and Posts.

Since the current institutional setup within the energy sector is dispersed among various ministries and agencies, policymakers have recognized the need to establish a coordinating unit in the Economic Planning Unit to streamline decision making. Moreover, because privatization of the utilities sector is a longer-term government objective, privatization of the National Electricity Board is under study. Since electric power is a toll good, some form of regulation will still be in force to ensure universal access. The privatization of Petronas is not now envisaged. The reason given is that oil is a common pool good, orderly exploitation of which can best be pursued by a public enterprise.

REGIONAL COOPERATION

Malaysia is highly dependent on the Asia-Pacific region for its trade in energy resources, the only exception being crude petroleum, which is imported mainly from Saudi Arabia and Kuwait. Most fuel products (with the exception of petroleum products) are imported from the Pacific region. From the ASEAN region, a substantial trade in petroleum products takes place between Malaysia and Singapore. Coal imports emanate from Indonesia and Australia, while coke imports are mainly from Japan and Australia. The total value of coal and coke imports was M$55 million in 1984. Out of an import value of M$3.2 billion for petroleum products in 1984, 38 percent was constituted by crude petroleum. Manufactured gas import value was M$33 million for the same year.

Crude petroleum exports are mainly to Japan, Singapore, South Korea, Thailand, the Philippines, and the United States, while LNG exports are entirely sold to Japan. Exports of Malaysian LNG to Japan began in 1983 under a 20-year contract. The first year's export volume amounted to 1.83 million metric tons. Subsequently, the export level increased from 3.7 million metric tons in 1984 to 4.5 million metric tons in 1985. In July 1985, Malaysia's first consignment of LPG, totaling 33,000 metric tons, from Petronas' gas processing plant at Kerteh, was shipped to Japan for delivery to Nippon Petroleum Gas Company, Daikyo Oil Company, and Taiyo Oil Company. Total exports of LPG in 1985 amounted to 196,000 metric tons. The major markets for the Malaysian crude were Singapore (33 percent), Japan (30 percent), and South Korea (12 percent) in 1984. However, exports to Singapore, the largest buyer since 1982, declined in 1985; Japan and South Korea significantly increased their imports, accounting for 34 percent and 16 percent, respectively. The largest buyer, Japan, imports about 52,000 b/d, while South Korea and Singapore average 32,000 b/d.

Energy cooperation within the ASEAN region has been a relatively recent development. The first attempt at regional cooperation on energy issues at the government level took place in 1975 with the establishment of the ASEAN Council on Petroleum, represented by the respective state oil companies. The ASEAN Petroleum Sharing Scheme, effected in 1977, called for collective action to assist any of its members in a global emergency situation of critical shortage or oversupply of petroleum or petroleum products. In 1979, the power utilities of Malaysia, Singapore, and Thailand formed the Joint Computer Applications Committee, which was later known as the Computer Applications Group upon the entry of the Philippines and Indonesia. It is a technical working group focusing on detailed technical issues and planning studies. In 1978, with energy increasingly viewed as a crucial economic issue, the Committee on Industries, Minerals and Energy placed nonpetroleum energy resources under its aegis. A working group under the auspices of the Committee on Science and Technology was formed to look into the development of nonconventional energy research, including solar, bio-energy, wind, and geothermal sources.

The highest authority on energy cooperation is the ASEAN Energy Ministers. A framework was formed for meetings at the senior official level to facilitate discussions on energy issues and to service the meetings of the ASEAN energy ministers. Following the first meeting of energy ministers in 1980, the ASEAN Expert Group on Coal was established. Areas delineated for cooperation included regional coal studies, joint market surveys, joint exploration, and survey activities. Another committee, acting as a forum for power utilities heads, was formed in 1981 and absorbed the existing Computer Applications Group. Among the areas of cooperation on power utilities are computer applications, rural and urban electrification, interconnections, research, and development. Informal cooperation through the ASEAN Chamber of Commerce and Industry forms another avenue for private sector exchanges on commercial matters.

On the basis of cooperation within ASEAN to ensure stable supply and cost of energy for long-term development, Malaysia is now in an advanced stage of negotiations with Singapore to supply 150 million cubic feet of piped natural gas per day, commencing in 1989 (*New Straits Times*, May 6, 1986). In the same vein, the Malaysian and Thai governments have designated an area of 7,102 square kilometers for joint exploration and exploitation of petroleum resources. The Malaysia-Thailand Joint Development Area is located roughly 145 kilometers northeast of Kota Bharu and approximately 282 kilometers east of Songkhla. Seismic interpretation, geological studies, and assessment of the potential of the area have been conducted. In a related development, an agreement was signed in March 1985 between Malaysia and Brunei covering the Fairley Baram field, which straddles the boundary between Sarawak and Brunei. Under the terms of the agreement, one-third of the production will accrue to Malaysia and the rest to Brunei (*Business Times*, March 23, 1985).

Since cooperation is a positive-sum game in which the gains to all exceed the sum total of individual losses, arrangements have to be made so that each party will gain from cooperation. However, perceptions of gains vary, and so does the acceptable time span within which such gains are realized. In Malaysia, it appears that regional interests are usually accorded priority only if they coincide with or promote national interests. Thus, increased cooperative efforts would necessarily imply the coincidence of needs as identified in the "energy plan" of Malaysia with those of the countries in the region.

NOTES

1. The nuclear option was deferred because priority was given to the development of other indigenous energy sources. However, research and personnel development will continue.
2. "Oil profit" refers to the balance of crude oil after the payment of royalty.
3. No allocation was made for the Bakun Project, estimated to cost M$12 billion, in the Fifth Malaysian Plan.

ELEVEN

Thailand's Energy Sector

BUNCHIRD KULPHANICH

Since the second oil price shock (1979–1980), the outlook for the international oil market has shifted. Oil-deficient countries have gone through a painful adaptation to higher oil prices. They have reduced consumption of oil products, increased oil supplies, and effected widespread substitution of indigenous for imported energy.

The sudden decline in the world oil price of around 40 percent during 1986 strongly affected the energy sector, leading to changes in the energy demand structure and in the formulation of energy policies.

ENERGY SECTOR PERFORMANCE

The energy sector plays a major part in the economic development of Thailand. There are several agencies related to energy planning development.[1]

In order to provide coherent energy policy, planning, and coordination, the National Energy Policy Committee was established, with the prime minister as chairman, the Cabinet ministers concerned and chairmen of certain state energy enterprises as members, and a person from the National Energy Policy Office as secretary. The duty of the committee is to plan the energy policy and to coordinate it with the overall economic policy plan.

Integration with Economic Planning and Policymaking

The National Economic and Social Development Board is in charge of the process of economic planning in its entirety. It is responsible for setting targets and ensuring that these targets are disseminated to the various components of the economic system. The board is also charged with setting up an organized framework for the implementation, coordination, and monitoring of an accepted plan. The process of energy planning is a major part of economic planning. It involves the preparation of quantitative estimates of the rate of expansion of the productive capacity of the economic system and the implications of grants for the energy sector, particularly in terms of investment requirements.

The relationships between the energy sector and other sectors of the economy from the demand side can be broadly summarized as follows:

- Energy consumption in industry
- Energy consumption in the household sector
- Energy consumption in the transportation sector
- Energy consumption in the agricultural sector

These linkages between the energy sector and each group of consuming sectors depend on the level of activity, technology, and price and income variables.

In Thailand, a large amount of funds is paid in support of various energy projects, especially petroleum and electricity, implemented by both the public and the private sectors. An energy policy suited to Thailand's economic condition is embodied in the energy plan draft, which is considered by the Energy Plan Steering Subcommittee and subsequently submitted to the National Energy Policy Committee.

As a result of the declining oil prices in 1986, the energy scene in Thailand has changed. Long-range plans for energy resource development, as well as energy pricing (particularly of oil and natural gas), need to be reviewed.

Declining oil prices caused a serious drop in oil/gas exploration and development because existing operators cut investment from previous years. The government passed on the benefits of declining oil prices to the consumer by reducing domestic oil retail prices. The price of natural gas was cut in response to declines in the price of fuel oil. In addition, the government reduced the electricity tariff rate as a result of the declining fuel price. To offset the cut in petroleum exploration and development investment by private firms, the government set policies to encourage the private sector to invest more in the energy sector. The 16-year-old Petroleum Act was updated in this context.

In the short and medium terms, the government is emphasizing energy conservation. The promotion of efficient use of energy in manufacturing industries and transport is the main objective. Table 11.1 indicates the long-term projections of energy supply and demand balance.

The Role of Noncommercial Energy Sources

Noncommercial energy (mostly firewood and charcoal) is very widely used for cooking and supplies some 90 percent of rural household energy. In the future, the share of noncommercial energy demand in total energy demand should

Table 11.1
Energy Supply and Demand Balance, 1986, 1991, and 2001
(million tons of crude oil equivalent)

	1986	1991	2001
Demand			
Petroleum product	13,226	15,573	—
Lignite and coal	1,894	3,227	—
Natural gas	4,037	6,884	—
Hydro power	1,349	1,580	—
Total[a]	20,506	27,264	40,904
Supply			
Domestic production	8,704	13,112	15,885
Import	11,802	14,152	25,019
Total	20,506	27,264	40,904

a. Excluding traditional energy demand.
Source: National Economic and Social Development Board, Bangkok, Thailand, 1987.

decrease because switching to LPG and electricity is taking place. However, there appears to be an impending serious firewood shortage. The demand is estimated to be substantially higher than sustainable domestic firewood supply.

EVIDENCE OF STRUCTURAL CHANGES IN ECONOMY AND ENERGY/ECONOMY RELATIONSHIPS

The Economic Effects of the 1973 and 1979 Oil Shocks

During the 1974–1979 period, after the first oil crisis, the government tried to insulate domestic oil consumers from the effects of the external oil price explosion by postponing domestic petroleum product price adjustments. It was not until 1980 and 1981 that the government seriously adjusted domestic prices. The economic effects of the two oil price increases are summarized below.

Effects on the Target Growth Rate of the Economy. An oil crisis affects the performance of every sector in the economy. Therefore, the target growth rate of the economy may not be achieved. As a result of the first oil crisis, the real growth rate of GDP dropped from 9.4 percent in 1973 to 5.4 percent in 1974. After the second oil crisis, the growth rate of GDP dropped from 6.3 percent in 1981 to 4.1 percent in 1982.

Effects on the Balance of Trade. As a result of the first oil crisis, the trade deficit increased from 3.996 billion baht in 1973 to 6.700 and 14 billion baht in 1974 and 1975, respectively. The trade deficit in 1974, the year following the first oil crisis, was alleviated by exporting surplus agricultural products at very high prices. After the second oil crisis, Thailand spent 32 percent of the foreign exchange earned from exports for importing oil, totaling approximately 62,000 million baht in 1981. The rising oil import bill was an important factor in Thailand's chronic trade account deficit that has endangered the country's financial position.

Effects on the Agricultural Sector. The burden of the oil crisis is not expected to be passed on to Thai farmers, except for the very few modernized large-scale farmers, because Thai farms are a long way from being fully mechanized. However, the increased price of fertilizer and increased cost of transportation adversely affected Thai farmers, reducing their incomes. The fishery sector, which requires large quantities of diesel oil as fuel, was particularly hard hit by the oil crisis.

Effects on the Industrial Sector. The immediate effect of the oil crisis was an inevitable increase in electricity rates, which had a widespread impact on the economy, particularly in the industrial sector. Raising the domestic price of oil directly affects the costs of production in industries such as cement, plastics, textiles, and construction.

Effects on the Transport Sector. It is obvious that transport touches all sectors of the economy. The impact of the oil crisis on transport will therefore affect all other sectors.

Effect on the General Price Level. As a result of the first oil crisis, growth of the consumer price index (CPI) and increased from 5 percent during the cheap oil era (1972) to 15 and 24 percent in 1973 and 1974, respectively. After the second oil crisis, CPI growth rose from 10 percent in 1979 to 20 percent in 1980.

Effects on Employment. Since manufacturing costs rose and demand for manufactured output fell during the resulting period of economic slowdown, employment problems occurred. Factories had to reduce their working hours and workers were laid off.

Economic Effects of the 1986 Oil Shock

In 1986, the price of oil dropped to around one-third of the 1981 price and world interest rates also fell. On the domestic front, the government introduced a number of policies designed to stimulate the economy, such as a decrease in the domestic oil price, a reduction in electricity charges, and a lower domestic interest rate. These changes all contributed to the rise of the Thai economy in the latter part of 1986. The main result in the first nine months of 1986 was a boom in manufactured exports, which was partly a delayed effect of the devaluation in 1984.

The corresponding upswing of the Thai economy in the latter half of 1986 was the percentage change of fuel consumption compared with the same period in 1985. In addition, as a side effect of the 1986 oil price drop, several oil companies decided to delay or to cut their petroleum exploration and development programs.

Government Polices Resulting from These Economic Effects

In 1972, Thailand's oil imports met 86 percent of total energy requirements. This reliance on imported oil decreased to 82, 61, and 50 percent in 1974, 1981, and 1984, respectively. It is clear that Thailand is still so dependent on oil imports that the country will inevitably be shaken by an oil price increase or disruption in the world supply situation.

Since 1973, the government has explored every possible means of securing a sufficient supply of oil for the country and a satisfactory means of payment. Therefore, efficient import substitution through expansion of domestic energy production is one of the specified tasks of the government. Even if domestic production could be increased, there would still be a substantial gap that must be closed by imports or even more efficient energy use. It is necessary for the government to implement effective price and nonprice policies for encouraging transportation, industry, and households to consume less and conserve more in order to achieve substantial savings.

Short-Term Measures. Short-term measures that the government used during the oil crisis to alleviate the impact on the economy included the following:

- Increasing awareness of the fact that the country needs the serious cooperation of the public in minimizing the use of oil
- Encouraging car pools as an oil-conserving measure
- Reducing the speed limit in order to get maximum mileage from each liter of gasoline consumed
- Limiting neon sign advertisements and lighted window displays in shops
- Limiting the number of hours during which gasoline can be purchased

Long-Term Measures. The oil crisis took the form of supply shortages and sharp increases in oil prices. Even though the recent fall in the price of oil has reduced economic tensions for oil-importing countries, no one can foresee the continuation of falling oil prices. It is clear that if economic development is the target of the country, Thailand will have to seek long-term solutions to energy problems. In order to solve the problems, the government has stipulated the following policies:

1. A nonprice policy to increase energy use efficiency by emphasizing conservation and reduction in the consumption of energy per unit of output, particularly in road transportation and industrial activities. Besides energy conservation in the transportation sector, it is necessary to restructure the transportation system into less-energy-consuming modes (by using more water and rail transportation). The communication and telecommunication systems will be expanded to reduce the need for travel; therefore,

traffic congestion in urban areas will be relieved. As for industrial use, the government will provide industrialists with fiscal and monetary incentives to improve the efficiency of their factories through the introduction of conservation measures and changes to less-energy-consuming production methods.

2. Speed up the development of indigenous sources of energy in order to reduce oil imports. Government agencies and the private sector will have to speed up the exploration for and development of various indigenous sources of energy, particularly the exploration for onshore and offshore oil, the development of natural gas in the Gulf of Thailand, exploration and development of hydro power resources and lignite deposits, experimentation with energy production from oil shale, exploration for uranium, and the development of other nonconventional forms of energy. Meanwhile, the government will develop an energy procurement and reserve system in order to ensure the country's future economic growth and stability.
3. Promote investment in energy activities by the public and private sectors, both domestic and foreign, and encourage cooperation with foreign countries in energy resource development, procurement of energy, and development of energy reserves.
4. Accelerate the production and use of all forms of energy at an appropriate level in rural areas. In addition, a consistent and coordinated plan will be formulated for energy procurement, production, and consumption, taking into account environmental conservation.
5. A price policy is an essential tool in most countries for increasing energy efficiency. At present, the government is adjusting energy pricing policies for all forms of energy to be timely, consistent, and appropriate. In order to avoid substitutions among different forms of energy due to distorted prices, as well as to provide justice between producers, sellers, and consumers, energy prices must reflect the true cost with no government subsidies.

After the second oil shock in 1979, energy consumption in the manufacturing and transport sectors clearly declined. This may be a result of conservation or of greater efficiency in the use of energy due to the increased oil price.

In Thailand, there were significant government investments between 1973 and 1986 in the power sector, accounting for the largest share of public investment in the energy sector.

CHANGING OIL PRICES AND ECONOMIC DEVELOPMENT

The energy prices set by the Thai government, particularly the prices of oil products, are somewhat distorted. The present distorted prices resulted from a series of regulated price shifts between 1973 and 1986. The regulated prices of oil products are as follows:

- Ex-refinery prices are based on the average posted price of the six refineries in Singapore. The import prices of directly imported products are based on the Singapore spot prices.

Table 11.2
Effects of Trade Patterns with Oil-Exporting Countries, 1984–86 (million baht)

		1984	1985	1986
Brunei	Export	237	388	388
	Import	5,165	8,971	4,716
China	Export	4,295	7,367	7,247
	Import	7,449	6,073	6,917
Indonesia	Export	1,101	1,176	1,512
	Import	1,868	1,657	1,642
Iran	Export	2,937	1,769	1,061
	Import	1	999	423
Malaysia	Export	8,287	9,646	10,007
	Import	11,921	14,825	10,118
Qatar	Export	82	109	187
	Import	273	153	1,534
Saudi Arabia	Export	4,599	4,382	5,358
	Import	20,865	7,121	2,522
Singapore	Export	14,722	15,350	20,577
	Import	19,373	18,746	15,845

Source: Bank of Thailand, *Quarterly Bulletin* 26, no. 3 (September 1986).

- Retail prices of various oil products are further distorted by wide differentials in their tax rates. These distortions widen the gap between retail prices of different oil products.

After the sharp decline in world oil prices in 1986, the government passed the benefit to the consumers by decreasing the domestic oil price three times, corresponding to the decreases in the world oil price.

The changing oil prices seem to have had a larger impact on Thailand's imports from oil-exporting countries than on Thailand's exports (Table 11.2). With the plunge in oil prices, the overall balance of trade improved in 1986.

CURRENT ECONOMIC DEVELOPMENT PLANNING AND STRATEGIES

Energy is a very important factor in promoting Thailand's economic and social development and stability. In Thailand, the total investment program in energy activities is very large in the private sector. Therefore, it is necessary to encourage the private sector to play a more active role in energy investment in order to help relieve the government's financial burden, as well as to improve the overall economic and energy situation.

The major government policies concerning current economic development can be summarized as follows:

- The effort to reduce loan rates was continued, in order to encourage financial institutions to follow suit.
- The excise tax on oil was increased, but only to the extent that it would not effect any change in retail oil prices.
- The oil reserve requirement was restructured.
- The price of cooking gas was fixed at an identical rate for the whole kingdom.
- Marketing margins of oil were adjusted upward.
- The price of natural gas sold to the Electricity Generating Authority of Thailand was reduced.
- Electricity charges were reduced.
- To liberalize energy trade and to make petroleum exploration and production more attractive for foreign investment, the 16-year-old Petroleum Act was updated by changing its terms and conditions.

CONCLUSION

The oil shocks caused economic slowdowns, reflecting Thailand's dependence on imported energy and on growth in the world economy as a whole. The effects of the oil shocks forced Thailand to stimulate indigenous energy resource development as well as to promote more efficient energy use. Appropriate policies of the government were introduced, with the hope that imported petroleum products would be replaced by indigenous products.

NOTE

1. These are the Ministry of Finance, Ministry of Industry, National Economic and Social Development Board, National Energy Administration, Department of Mineral Resources, Electricity Generating Authority of Thailand, Metropolitan Electricity Authority, Provincial Electricity Authority, Petroleum Authority of Thailand, and Royal Forestry Department.

TWELVE

Energy and the Economy in the Philippines

CAMILO JOSE LIM, JR.

ENERGY ORGANIZATION

Structure and Functions

Prior to the first oil shock of 1973–1974, energy planning and management in the Philippines were made difficult by the presence of antiquated institutional laws and various multinational interests. Comprehensive planning and management of the sector came about only with the creation of the Department of Energy in 1977.

The Department of Energy, which was transformed into a ministry shortly thereafter, was charged with the task of rationalizing and integrating energy policies and programs. It was responsible for the formulation and implementation of policies, plans and programs for the development of indigenous energy resources. In addition, it monitored and supervised international trade, domestic refining, marketing, and distribution of petroleum products; regulated and controlled private sector involvement in the energy sector; coordinated the electric power industry; and supervised and coordinated research on nonconventional energy sources.

The ministry had two line bureaus—the Bureau of Energy Development and the Bureau of Energy Utilization. The Bureau of Energy Development was responsible for the promotion of all indigenous energy resource development, except hydro and nonconventional, and for the regulation of private domestic and foreign firms involved in such activities. The Bureau of Energy Utilization, on the other hand, was responsible for regulating the import, refining, marketing, and distribution of crude oil and petroleum products; the preparation and implementation of energy conservation programs and utilization standards; the collection and interpretation of energy sector data; and the formulation of energy demand and supply forecasts.

The ministry also exercised control over three other agencies:

- The Philippine Atomic Energy Commission, responsible for the planning and regulation of the country's nuclear energy program
- The Board of Energy, a quasi-judicial institution that primarily regulated petroleum product prices and retail tariffs for electric power supplied by private utilities
- The Center for Nonconventional Energy Development, whose main responsibility was to encourage research on the application of new energy technologies.

Closely linked to the ministry for policy guidance, but not formally responsible to it, were two large government corporations whose operations largely affect the government's energy program. These were the National Power Corporation (NPC) and the Philippine National Oil Company (PNOC). The NPC was primarily responsible for the generation and transmission of electricity; the PNOC was the government's main executing agency in the petroleum sector, although it was also active in coal and geothermal resource development.

There were other agencies in the energy sector whose activities were not closely coordinated with those of the Ministry of Energy: the National Electrification Administration (NEA), the Philippine National Alcohol Commission (PNAC), and the National Coal Authority (NCA). The NEA was responsible for implementing the government's rural electrification program through the electric cooperatives. Unlike NPC and PNOC, the NEA reported not to the Ministry of Energy but to the Office of the President. The PNAC was established to provide overall policy guidance and support to the government's ambitious alcogas program, which sought to replace imported gasoline in the transport sector with a blend of gasoline and alcohol domestically produced from sugar and other crops. The NCA was the most recently created body responsible for promoting the development of the domestic coal industry.

Recent Developments

With the pending reorganization of the whole government machinery, the Ministry of Energy was transformed into the Office of Energy Affairs (OEA) and was placed under the Office of the President. Although kept intact, the line bureaus had their personnel reduced. Many of the technical staff who were seconded by PNOC to the Ministry of Energy were either recalled or asked, if they wished, to fill the regular positions of OEA.

There is a plan ultimately to constitute the energy sector agencies under the Department of Natural Resources. If this plan is carried out, the former Ministry of Energy would be transformed into one of the line bureaus of the newly created Department of Energy, Environment, and Natural Resources.

Linkage with Economic Planning and Policymaking

At present, energy policies and programs are formulated in conformity with the basic Energy Development Program of the Ministry of Energy (OEA), which, in turn, forms an integral part of the National Development Plan. To ensure consistency with other sectoral policies and programs, the minister of energy (officer in charge) sits as a member of the National Economic and Development Authority (NEDA) board. The NEDA board, which is chaired by the president and composed of the various ministers, is the foremost planning and policy-making body in the country.

POLICY RESPONSES

The First Oil Shock, 1973–1974

Encouraged partly by the cheap oil prices of the 1950s and 1960s, the Philippines embarked on an import-substituting industrialization policy that relied heavily on imported materials and capital equipment. It was eventually observed that despite the labor surplus of the Philippine economy, the share of labor-intensive manufactured goods in the country's total exports had been very small. The postwar period up to the early 1970s also witnessed the growth of energy-intensive industries such as textiles, paper and paper products, chemicals and chemical products, rubber products, basic metals and metallic products, and machinery. Thus, when the first oil shock hit the country, it was heavily dependent on imported oil to the extent of over 90 percent of total commercial energy consumed.

Despite the relatively favorable prices of exports at that time and the significant increase in exports brought about by the peso devaluation and the enactment of the Export Incentives Act in 1970, the quadrupling of oil prices resulted in a trade deficit of $418 million in 1974, which grew to about $1.3 billion in 1978. From 11.7 percent in 1973, oil imports rose to about 20 percent of total imports and easily took a third of total Philippine exports proceeds. Consequently, the current account balance registered huge deficits, which were partially offset by inflows of foreign capital. In particular, short-term capital inflows, which were primarily used for financing oil imports, almost doubled, from $657 million in 1973 to $1.138 billion in 1974, and steadily increased thereafter. The level of external debt rose from $2.9 billion (27.2 percent of GNP) in 1973 to $13.4 billion (45.3 percent of GNP) in 1979, representing an average growth of 29 percent over the six-year period.

Domestically, inflation surged to 34.1 percent in 1974, despite efforts by the monetary authorities to restrain credit and money supply expansion. However, subsequently, inflation stabilized at single-digit levels and the economy grew at a respectable rate of 5–8 percent through a countercyclical fiscal policy and a cautious monetary stance made possible by massive inflows of medium- and long-term loans.

One of the direct responses of the government to the energy problem was the creation of PNOC in 1973. Previously, oil multinationals controlled the entire supply of oil to the Philippines. With the creation of PNOC, the government became actively involved not only in the supply of petroleum but also in the exploration for and development of domestic petroleum resources. The PNOC is now engaged in all aspects of the oil industry, from exploration and development to transport, refining, storage, and marketing.

To encourage the development of domestic energy resources and reduce the Philippine's heavy dependence on imported oil, the old system of oil exploration through concessions was changed to the service contract scheme. The enactment of Presidential Decree no. 87, better known as the Oil Exploration and Development Act, attracted foreign as well as local prospectors in the search for oil in the country.

Presidential Decrees no. 972 (Coal Development Act) and no. 1141 (Geothermal Development Act) were issued in 1976 and 1978, respectively, to encourage the development of alternative energy resources. The coal development program succeeded in stimulating production from only 39,000 metric tons in 1973 to 254,515 metric tons in 1978, while the geothermal program succeeded in establishing the Philippines as the world's second largest geothermal producer after the United States.

In summary, the Philippines' response to the first oil shock was a direct reaction to the problems at hand—securing energy supply, establishing the appropriate institutional mechanisms and legal requirements, and direct market intervention through market allocation and rationing during the period of tight supply. Even with respect to pricing, the government did not have the resources to shield domestic consumers entirely from external price increases. However, much care was exercised to protect the consumers, especially in metro Manila (the seat of government), from bearing the full brunt of price increases. While this seemed to be the most logical and politically acceptable policy at the time, distortions in consumption and other negative effects began to surface during the period of the second oil shock.

The Second Oil Shock, 1979–1980

The Philippines was just completing its adjustment to the oil crisis of 1973–1974 when the second oil shock struck. There was a general expectation that oil prices would continue to rise; therefore the objectives of energy security, energy conservation, and reduced dependence on oil continued to be emphasized.

The second oil shock served to intensify and improve the policies that had been adopted in response to the first. Increases in international oil prices were fully and promptly passed on to consumers, although by altering the tax component of the final price, the government slowly restructured fuel prices to correct the refinery imbalance brought about by discriminatory pricing in the past. Fuel switching and energy conservation were encouraged through various financial and fiscal incentives. The manufacture and use of energy conservation and substitution projects were included in the Board of Investments' annual Investment Priorities Plan.

Earlier efforts to develop and diversify domestic energy supply had begun to bear fruit. Commercial production of oil and geothermal energy started in 1979, and the country's dependence on imported oil decreased to 72 percent.

The macroeconomic impact of the second oil shock, however, was more severe. In 1980 alone, the country's oil bill reached $2.2 billion, accounting for 28.8 percent of total imports and claiming 38.5 percent of total export proceeds. Starting in 1979, the country's external payments position went in the red and further deteriorated as a result of weak export prices and high interest payments.

The country's foreign liabilities had reached a level of $13.4 billion in 1979 and increased to $20.9 billion in 1981. The growth of the economy's output (GDP) continued to decelerate from 6.2 percent in 1979 to 3.9 percent in 1981, while domestic inflation reached double-digit levels despite fiscal and monetary restraint.

To summarize, the second oil shock had far more extensive and permanent effects on the economy than the first. While relative success was attained in the energy sector in terms of reducing dependence on imported oil, several factors contributed to the overall deterioration of the Philippine economy. First, the Philippines had not fully adjusted to the impact of the first oil crisis. Second, the Philippines opted for a "growth through external debt" policy prior to the second oil shock that succeeded in stabilizing domestic inflation and artificially promoting growth, but proved difficult to sustain. Finally, the external environment (weak export prices, restrictive capital markets, and increasing protectionism) presented further obstacles to growth.

The Period of Decline, 1982–1986

The decline in oil prices, from 1982 to mid–1986, provided much needed relief to the Philippines. The country's oil bill declined from a record high of $2.439 billion in 1981 (30.7 percent of imports and 42.6 percent of exports) to $827 million in 1986. However, the cumulative impact of previous economic policies, together with adverse external conditions, more than offset the favorable effect of the oil price decline. In 1983, the balance-of- payments deficit reached crisis proportions, prompting the government to institute exchange controls and seek a restructuring of its foreign obligations.

By then, the Philippines had succumbed to stagflation-recession with high inflation. The growth in economic output (GDP) slipped further, from 2.9 percent in 1982 to 0.9 percent in 1983; and for two consecutive years, 1984 and 1985, the economy posted negative growth rates of 6.0 percent and 4.4 percent, respectively. Inflation reached double-digit levels and soared to 50.3 percent in 1984.

The decline in the world price of oil could not be immediately translated into lower domestic petroleum prices because of a rapidly deteriorating exchange rate. The equivalent peso cost of oil continued to increase even while global oil prices collapsed. In fact, in 1983 and 1984, a series of domestic oil price hikes were implemented. It was only when the peso restabilized during the first quarter of 1985 that domestic petroleum prices were adjusted downward.

The decline in oil prices presented the government with an opportunity to further rectify the refinery imbalance caused by the wide price differentials between substitute products and to restructure the various components of domestic petroleum prices for revenue generation and price stabilization. An ad valorem tax on petroleum products was introduced in 1984, and the Oil Price Stabilization Fund replaced the Crude Oil Equalization Fund as a mechanism to stabilize domestic petroleum prices.

Electricity rates were adjusted to correspond to the lowering of domestic petroleum prices. A program to reduce the subsidy on the electricity consumption of residential and commercial customers in metro Manila also was approved in 1984. The program entails the gradual reduction of the subsidized consumption of residential and small commercial customers to 50 kilowatt-hours by 1990.

Meanwhile, domestic energy development and diversification continued. In 1985, the country's dependence on imported oil was reduced to about half (51 percent) of its total energy requirements. Fuel switching was further encouraged with the launching of a cement industry rationalization program that aimed at converting the cement plants from oil to coal.

EVIDENCE OF STRUCTURAL CHANGES IN THE ECONOMY

Energy-GDP Relationship

The literature on energy-economy interaction is rife with evidence linking per capita energy consumption to per capita GDP. This relationship is evident in the Philippines, where per capita energy consumption grew with real per capita output. From 1970 to 1980, per capita energy consumption increased from 1.48 barrels of fuel oil equivalent to 2.0. Correspondingly, per capita real GDP increased from 1,377 pesos in 1970 to 1,917 pesos in 1980. In later years, when real per capita GDP started to fall, per capita energy consumption also declined.

Studies have been made on the elasticities of energy demand to income. The results show that aggregate demand, particularly for conventional energy, is quite responsive to changes in economic output. However, income elasticity becomes less when more recent data are used. (See Table 12.1)

Table 12.1
Summary of Existing Econometric Estimates of Income Elasticity of Aggregate Energy Demand

Source	Period covered	Energy definition	Income elasticity estimate
Ministry of Energy (1980)	1962–79	Conventional	1.67
		Petroleum	1.73
Alejo (1983)	1970–79	Conventional	0.91 (1979)
		Petroleum	0.97 (1979)
Alejo (1986)	1974–84	Total primary	
		Short-run	0.55
		Long-run	0.87

Sources: Alejo, Leander, "A Study of Energy-Economy Interaction in the Philippines," Monograph Series No. 3 (Philippines Institute for Development Studies, 1983); "Energy Demand Management Policies in the Philippines," Energy Study Series No. 9 (Asian Development Bank, 1986); Ministry of Energy, "Ten-year Energy Program, 1980–1989" (Manila, 1980).

Energy and Oil Intensities

As indicators of efficient energy use, average energy and oil intensities were computed for the period 1970–1986. The results show that overall efficiency of energy use has increased, as indicated by a decrease in the amount of energy use per unit of output from 1,315 barrels of fuel oil equivalent/million pesos in 1970 to 151/million pesos in 1986. (See Table 12.2). The decline in energy intensity was particularly rapid after the first and second oil shocks. Because of the increased contribution of domestic energy resources other than oil, average oil intensity also declined from 1,241 barrels of fuel oil equivalent/million pesos in 1970 to 84/million pesos in 1986.

Between 1974 and 1984, however, transformation losses also increased. From a level of 23 percent of total primary energy requirements in 1974, transformation losses increased to about 37 percent in 1984. This was brought about by the rise in the proportion of energy consumed in the form of electricity, from 11.7 percent of final energy consumption in 1974 to 20.1 percent a decade later, and a general deterioration in the condition of transmission and distribution systems. Another factor that has contributed to the increase in transformation losses is the rise in losses due to pilferage. In the metro Manila franchise area alone, losses due to theft have been reported to reach as high as 11 percent. The incidence of losses due to pilferage in areas outside metro Manila is higher.

Table 12.2
Energy and Oil Intensities, 1970–86
(barrels of fuel oil equivalent per million pesos)

Year	Energy intensity	Change (%)	Oil intensity	Change (%)
1970	1,315		1,241	
1971	1,226	-6.8	1,154	-7.0
1972	1,130	-7.8	1,059	-8.2
1973	882	-21.9	888	-16.1
1974	756	-14.3	605	-31.9
1975	692	-8.5	566	-6.4
1976	620	-10.4	494	-12.7
1977	582	-6.2	471	-4.7
1978	530	-8.9	424	-10.0
1979	448	-15.5	357	-15.8
1980	366	-18.3	276	-22.7
1981	306	-16.4	224	-18.8
1982	281	-8.3	201	-10.3
1983	256	-8.6	178	-11.4
1984	176	-31.5	104	-41.6
1985	151	-14.0	82	-21.2
1986	151	0.0	84	2.4

Source: Office of Energy Affairs and National Economic and Development Authority, *Philippines Statistical Yearbook* (Manila, 1987).

Sectoral Consumption Patterns

Energy consumption in the Philippines is classified into six sectors: industry, transportation, commercial, residential, agricultural, and others. Sectoral classification has been a problem because only sectoral consumption of electricity is officially reported (because of facility in monitoring) and because data, especially for noncommercial energy, are not readily available. Moreover, the energy consumption of some sectors is understated due to lack of data and inappropriate classifications. For instance, the agriculture sector is known to consume significant amounts of commercial energy for agricultural equipment and fishing craft, and a sizable amount of noncommercial energy. However, some is recorded as part of industry or transportation. Nevertheless, these data give broad indications of sectoral energy demand.

The structure of energy consumption by sector in the Philippines changed very little over the period 1974–1984. Industry still consumes about half of total energy, while the commercial sector's share of total energy consumption has remained flat at about 6 percent. (See Table 12.3). The only noticeable changes were in the residential and transportation sectors. With higher living standards and an aggressive rural electrification program. the residential sector expanded its share of total energy consumption from 10.9 percent in 1974 to 14.3 percent in 1984. The transportation sector gradually reduced its share from 28.6 percent in 1974 to 24.2 percent in 1984. The reduced share of the transportation sector in total energy consumption could be attributed to the increased cost of fuels brought about by successive increases in domestic petroleum prices and more efficient energy use due to increased fuel efficiency of vehicles, better fleet management, and increased energy consciousness among car owners. Furthermore, there was a shift in the mode of transportation, from water and air to land, indicating an increased share of value added in land transportation.

A comparison of sectoral energy intensities shows a much higher and more rapid reduction in energy intensity in the transportation sector over the period 1974–1984. Industry displayed a somewhat lagged response to oil price increases due to structural rigidity in the sector.

OUTLOOK AND STRATEGIES

The new government has come up with the Medium-term Philippine Development Plan for the period 1987–1992. GDP is targeted to grow at an average yearly rate of 6.9 percent, with industry providing the major boost. To support this growth in economic output, total energy supplies are expected to grow by an average of 4.1 percent annually over the same period.

The use of energy in the form of electricity would continue to increase to about 44 percent of total energy supplies by 1992. However, oil's share in the generation of electricity would decrease to 31 percent. No new oil plants are scheduled for construction over the period 1987–1992, since the bulk of increase in installed generation capacities would be provided by coal and geothermal plants.[1]

The perception among government and energy sector planners is that oil prices will remain soft for the rest of the decade and will tend to rise again in the 1990s. The uncertainty that surrounds the market for oil, however, warrants a more conservative stance in energy planning.

Over the medium term, the Philippines will continue to aim for security in energy supplies and reduced dependence on imported oil. At the same time, efficient energy use will be promoted through pricing and other tools. The following strategies were formulated to support the above objectives:

Supply and Resource Development

- Diversification of the geographical sources of petroleum supply for increased reliability and security, including the maintenance of domestic oil and gas exploration and development programs

Table 12.3
Energy Consumption by Sector, 1974–84 (mmtoe)

Sector	1974	1975	1976	1977	1978	1979	1980	1981	1982	1983	1984
Industry	4.52	4.82	5.25	5.65	5.43	5.72	5.68	5.55	5.28	5.82	5.47
Transportation	2.68	2.88	2.89	2.98	3.08	3.14	2.85	2.59	2.67	2.62	2.64
Commercial	0.61	0.67	0.72	0.75	0.69	0.74	0.69	0.71	0.74	0.75	0.69
Residential	1.02	1.13	1.24	1.29	1.65	1.69	1.72	1.59	1.68	1.75	1.56
Agriculture	0.28	0.26	0.26	0.28	0.30	0.33	0.28	0.26	0.24	0.27	0.23
Others	0.25	0.28	0.28	0.26	0.31	0.34	0.35	0.35	0.35	0.39	0.31
Total	9.36	10.04	10.64	11.21	11.46	11.96	11.57	11.05	10.96	11.60	10.90
Percent Share											
Industry	48.29	48.01	49.34	50.40	47.38	47.83	49.09	50.23	48.18	50.17	50.18
Transportation	28.63	28.69	27.16	26.58	26.88	26.25	24.63	23.44	24.36	22.59	24.22
Commercial	6.52	6.67	6.77	6.69	6.02	6.19	5.96	6.43	6.75	6.47	6.33
Residential	10.90	11.25	11.65	11.51	14.40	14.13	14.87	14.39	15.33	15.09	14.31
Agriculture	2.99	2.59	2.44	2.50	2.62	2.76	2.42	2.35	2.19	2.33	2.11
Others	2.67	2.79	2.63	2.32	2.71	2.84	3.03	3.17	3.19	3.36	2.84
Total	100.00	100.00	100.00	100.00	100.00	100.00	100.00	100.00	100.00	100.00	100.00
Growth Rate (%)											
Industry		6.64	8.92	7.62	-3.89	5.34	-0.70	-2.29	-4.86	10.23	-6.01
Transportation		7.46	0.35	3.11	3.36	1.95	-9.24	-9.12	3.09	-1.87	0.76
Commercial		9.84	7.46	4.17	-8.00	7.25	-6.76	2.90	4.23	1.35	-8.00
Residential		10.78	9.73	4.03	27.91	2.42	1.78	-7.56	5.66	4.17	-10.86
Agriculture		-7.14	0.00	7.69	7.14	10.00	-15.15	-7.14	-7.69	12.50	-14.81
Others		12.00	0.00	-7.14	19.23	9.68	2.94	0.00	0.00	11.43	-20.51
Total		7.26	5.98	5.36	2.23	4.36	-3.26	-4.49	-0.81	5.84	-6.03

mmtoe = Million tons of oil equivalent.

Source: Energy Balance, Office of Energy Affairs, Manila, 1987.

- Accelerated diversification to alternative sources of energy, with emphasis on indigenously abundant and regenerative forms
- Establishment of a logistical support infrastructure to enable efficient handling, storage, distribution, and marketing of traditional and new energy resources
- Widespread application of appropriate decentralized energy technologies for rural areas and for inaccessible areas
- Establishment of an adequate stockpile of energy as a protective measure against possible supply disruptions

Energy Demand Management

- The use of fiscal measures for reflecting the real economic and social costs of energy
- Direct government management of demand levels in time of emergency
- Direct government intervention to accelerate the development of certain markets for nonoil fuels
- Institutionalization of energy conservation

The energy program in the medium term will require investments amounting to 88.7 billion pesos. A large part of this (80.5 percent) will still come from the public sector, although it is government policy to invite increasing amounts of private capital in energy projects. Legislation has been passed encouraging private sector participation in the capital-intensive power generation subsector. Because of financial constraints, the government is currently giving priority to the completion of energy projects that are now in the "committed" stage of implementation. The rehabilitation and restoration, as well as the improvement and upgrading, of existing facilities are given preference over replacement or new construction.

CONCLUSION

Since the mid–1970s, the Philippines has undergone major economic and political changes. The changing oil scenario has had as much to do with these changes as have the economic policies that were pursued.

Oil prices have been, and will continue to be, an important factor in the economic plans of countries like the Philippines. Its importance underscores the need to remain vigilant in the face of declining oil prices and to speed up the development of domestic energy resources to effectively insulate the economy from future oil shocks.

The Philippines stands a good chance of further reducing its dependence on imported oil to less than half (47 percent) of its total energy requirements by 1992. With major energy projects utilizing domestic energy resources in the pipeline, the contribution of local energy resources to total energy supply is expected to increase from 45 percent in 1987 to 52 percent by 1992.

Nonconventional energy sources such as agricultural wastes and bagasse, a by-product of sugarcane, would increase its share to about 41 percent of domestic energy supplies by 1992.

There is still much room for expanding the contribution of domestic energy resources to the total energy requirements of the economy. However, this will require the early resolution of issues regarding pricing, particularly for geothermal steam and electricity. In the pricing of electricity, petroleum, and other energy products, great caution should be exercised to avoid distortions in consumption and possible adverse consequences on the environment.

NOTE

1. Most recent estimates of power demand reveal a possible shortfall by 1989–1990 that, because of time constraints, may require the setting up of gas turbines (300 MW) or the rehabilitation of existing oil plants to service peak requirements.

THIRTEEN

The Energy Outlook in Indonesia

HENDRO PRAWOTO

INTRODUCTION

In view of the close interrelationships between economic and population growth and the supply of energy, energy policy forms an integral part of national development.

It is already apparent that at the present level of global economic activity, the problems of energy needs and availability will soon reach a critical point and will have a negative impact on economic growth.

Numerous countries are endeavoring to free themselves from oil dependency. The increased demand for energy also has driven individual countries to seek greater efficiency in energy use.

A fully coordinated and comprehensive policy on energy is required for Indonesia. A densely populated country with low per capita income, it requires rapid economic growth and development. Rapid growth and industrialization will result in increased energy consumption. Indonesia has to put aside a certain share of its oil and natural gas production for export to earn seriously needed foreign exchange for development. Exports of petroleum and liquified natural gas (LNG) are key sources of foreign exchange and Indonesia's main sources of internal government revenues that are necessary for financing development. Yet, Indonesia also requires more energy to meet its domestic demand, which is rapidly increasing every year in line with its economic growth and rising population.

The maximum utilization of nonoil and gas energy sources available in Indonesia, such as coal, hydro, and geothermal power, appears to be the answer to the problem. New and renewable energy sources, such as wind and solar energy, have a role to play in this effort as well.

Energy policy should go hand in hand with overall development strategy in accordance with the aspirations of the people, as expressed by the 1945 constitution. Policy formulation should be based on the principle that air, soil, and natural wealth, controlled and owned by the state, should be fully utilized for the welfare of the general public.

Energy development and production is influenced by several important factors, including the type of energy resource available, supply and demand conditions, and the prevailing technology. The management of these factors will determine the cost of production and services; this, with demand, in turn will decide the market price.

The Republic of Indonesia is an archipelago of about 13,000 islands with a total land area of about 2 million square kilometers. The country has a population of 165 million. The average density is 86 inhabitants per square kilometer, but the distribution varies from 3 to 761 inhabitants per square kilometer. The main economic activity of the majority of people is agriculture, but this sector accounts for only about 30 percent of the gross domestic product.

During the period 1960–1970, energy consumption grew at an annual rate of 7.4 percent—during the second REPELITA (Five-Year Plan), 1975–1979, the average annual consumption growth rate reached 13.8 percent, but fell to 5.6 percent during the third REPELITA, 1980–1984.

OIL DEPENDENCY

Dependency on oil has made the government of Indonesia aware that guidelines to manage domestic energy supply and demand are necessary so that an adequate share of total oil and gas production can be directed to export markets. This concern has guided energy policy.

Since other sources of government revenues are presently lacking, revenues generated from oil and other natural resources are essential to fund national development efforts. The efficient use of such revenues has been necessitated by the growing capital expenditures called for in successive five-year plans.

ENERGY POLICY

Various energy policy measures, including what we refer to as intensification, diversification, conservation, and indexation, have been adopted in order to assist planning. These measures are explained as follows:

- Intensification—To increase exploration and survey activities, a more exact understanding of the production potential of various energy sources in economically meeting the demand for energy must be obtained.
- Diversification—To reduce the dependency on petroleum, efforts are being made to develop substitute energies to fulfill internal demand.

Table 13.1
Increase of Energy Sources During Second, Third, and Fourth REPELITAs (thousand barrels of oil equivalent)

Energy source	Realization at end of second REPELITA	Realization at end of third REPELITA	Realization at end of fourth REPELITA
Natural gas	23,495	43,310	55,248
Coal	647	141	28,244
Hydro	3,852	11,636	24,330
Geothermal		418	1,958
Nonoil subtotal	27,994	55,505	109,780
Crude oil	131,009	167,631	182,408
Commercial energy total	159,003	223,136	292,188

Source: BAPPENAS, National Development Planning Agency, Republic of Indonesia, 1987.

- Conservation—To achieve more efficient energy use, conservation principles should be applied by all development operators at each stage of decision making. All consumers should become fully aware of the scarcity of energy and be encouraged to adopt a "save energy" way of life.
- Indexation—A mix or type of energy that is most appropriate for each sector or activity must be determined.

THE AIM OF A NATIONAL ENERGY POLICY

The objectives of a national energy policy are the following:

- To make energy available domestically
- To make energy available for export
- To economize in the use of oil as fuel
- To develop new energy sources as oil substitutes
- To maintain and preserve the natural environment
- To increase national resilience.

TARGET OF THE FOURTH FIVE-YEAR NATIONAL DEVELOPMENT PLAN, 1985–1989

The target set by the fourth REPELITA is to continue efforts to decrease the role of oil as a fuel and increase the role of other energy sources. Locally, oil's share in commercial energy consumption is targeted to decrease from 75 percent in 1983–1984 to 62 percent by 1988–1989 (Table 13.1).

Table 13.2
Long-Term Estimates of Commercial Energy Demand, 1990–2001

	1990–1991	2000–2001
GDP growth (%)	5	5.0–6.5
Population growth (%)	2	1.8–2.0
Income elasticity	1.63	1.3
Low estimate	332	630
High estimate	332	710

Source: BAPPENAS, National Development Planning Agency, Republic of Indonesia, 1987.

The development of noncommercial energy (such as biomass and agricultural waste) needs to be directed toward filling the shortage of firewood in rural areas—these sources are estimated to be capable of meeting 40 percent of total energy consumption (commercial and noncommercial) by the end of the fourth REPELITA.

The National Energy Coordinating Board was established to institute a coordinated energy policy. This board is chaired by the minister of mining and energy, and its members include ministers and directors general whose departments are directly involved in the energy sector, as well as those doing research on energy-related issues.

LONG-TERM DEMAND ESTIMATES

Long-term forecasts of commercial energy demand are based on several crucial parameters, particularly growth of GDP and population. Based on assumptions regarding these, estimates can be reached that are illustrated in Table 13.2.

INFRASTRUCTURE

The following infrastructure is needed to support energy policy:

- Research and development, including developing scientific skills and research work, developing infrastructure for scientific research (institutes, labs, development centers), providing research funds, and developing scientific ability in specific energy fields
- Energy industry, including promoting national economic growth by increasing the internal capacity to supply energy as an integral part of energy policy; it includes all industry supporting the development of energy supply as a whole, exploration and discovery, extraction, manufacturing, marketing, and transportation)
- Energy pricing and investment climate—the pricing of energy is influenced by the government, depending on energy source, economic growth, and a wide range of strategic objectives; it is necessary to create a healthy investment climate for engaging

in energy activities that will benefit the business sector, the government, and the consuming public.

ENERGY CONSUMPTION IN INDONESIA

In 1986–1987, primary energy consumption in Indonesia was estimated at about 252 million barrels of oil equivalent. The share of oil in total commercial energy consumption decreased from 70 percent in 1985–1986 to 65 percent in 1986–1987.

The pattern of energy consumption shifted between 1983–1984 and 1986–1987. Coal and hydropower were most significant in substituting for oil, though there were also slight increases in shares of geothermal and gas.

Coal

Indonesian coal resources are estimated to be as much as 25 billion tons, of which about 3 billion tons currently constitute proven reserves. The reserves are located mainly in east Kalimantan and south Sumatra. Coal production increased to about 2 million tons in 1985, of which about 1 million were exported. The government's policy is to achieve maximum utilization of domestic coal supply in power plants and the industrial sector. Development of fields such as those at Bukit Asam and Muara Tiga has started in line with this objective. In 1981, the government entered into a number of production-sharing agreements with private foreign and Indonesian joint-venture consortia for exploration and exploitation of Kalimantan coal areas.

The long-run marginal cost of Kalimantan and Sumatra coal is about $23/ton (6,000 Kcal/kg) delivered to Java. Domestically produced coal is competitive with imported coal at its current price.

Geothermal

Indonesia has several geothermal energy resources. Since geothermal steam provides an alternative source of energy, its price should be determined in line with the energy diversification policy. The price of steam is determined on the basis of its fuel oil equivalent price, which currently is not very attractive.

After the decline in oil prices in 1986, the further development of geothermal resources was not competitive with coal or even fuel oil. Indonesia's total geothermal reserves (about 10,000 Mw) are large and well distributed (5,500 Mw on Java). Geothermal production accounts for only about 0.67 percent of total commercial energy consumption from two sites in operation—Kamojang and Dieng. Other sites have been investigated, and some (Rawa Danau, Banten; Buyan Bratan, Bali; and Lahendong, north Sulawesi) seem to have potential if prices become more favorable.

Natural Gas

The remaining proven reserves of natural gas, mostly located offshore in the South China Sea as well as Kalimantan and Sumatra, have been estimated at about 80 trillion cubic feet (ft^3) (1.9 billion tons of oil equivalent; toe). About 90 percent of these reserves are nonassociated gas. The production of natural gas has rapidly increased from 1 billion ft^3 (24 million toe) in 1980 to 1.6 billion ft^3 (39 million toe) in 1986. Export of LNG, which began in 1978, grew from 0.5 billion ft^3 (12 million toe) in 1980 to 0.8 billion ft^3 in 1985, and LNG's share of total export earnings rose from 8 percent to 17 percent over the same period. The local utilization of natural gas accounts for 20 percent of total commercial energy consumption.

For domestic markets, which are concentrated in Java, the government has promoted the development of gas fields in and around Java with an estimated potential of 100–180 billion ft^3 per year. To achieve this potential, the government should fix product gas prices to encourage consumer usage.

Hydro Power

Indonesia has considerable hydro resources, estimated at about 75,000 Mw, but development to date has been limited by geographic distribution relative to the demand for electricity. The greatest potential (over 35 percent) lies in Irian Jaya, which accounts for less than 1 percent of total electricity demand. In contrast, Java has less than 10 percent of total hydro potential but accounts for 80 percent of current electricity consumption. In fact, in Java, about 2,000 Mw of hydro power have already been developed or are under construction.

Oil and Petroleum Products

The proven remaining oil reserves total about 9.5 billion barrels (1.3 billion toe). Current oil production is about 1.3 million barrels per day (b/d), down from 1.6 million b/d in 1980, as a result of OPEC production quotas (set at 1.2 million b/d in 1986 for Indonesia).

Since the introduction of the first REPELITA in 1968, energy consumption in Indonesia has grown rapidly (nearly 10 percent annually), and about 70–80 percent of consumption has been accounted for by oil. It is necessary to change attitudes and to diversify energy sources to reduce oil dependency. The consumption of petroleum products has grown rapidly with total commercial energy consumption. Over the past decade, the rapid development of industry and transportation has figured prominently in the rising consumption of petroleum products.

ENERGY FOR FUTURE DOMESTIC DEMAND

Future domestic energy demand will be satisfied by local sources, and oil dependency will decrease. The consumption of energy in Indonesia is still largely dependent on oil, but the role of oil is decreasing. On the other hand, the role of coal and natural gas will increase considerably. The development of hydro and geothermal will also accelerate with the expansion of electric power.

Until 1981, the high growth rates of consumption in subsidized products such as kerosene and the increase of crude oil prices resulted in a significant increase in national budgetary subsidies for petroleum products. This continued even though a petroleum product price adjustment was introduced in 1984. The growth of coal consumption in the industrial sector was partially due to the government's policy of diversification.

The role of pricing as a policy instrument has been very effective. In the past, the prices of petroleum products were set with particular attention given to distributional concerns. There was also a view that diesel fuel prices should be kept low, particularly to accelerate industrial growth. This policy changed as concerns about energy conservation and more efficient use of budgetary resources increased.

GROSS DOMESTIC PRODUCT AND POPULATION

The growth rate of GDP in Indonesia ranged between 5.8 and 8.5 percent annually from 1976 to 1982. It decreased thereafter as a result of declining crude export prices. Until 1995, the growth rate of GDP is projected at about 2.5–3.4 percent annually.

The population growth rate in Indonesia is 2.2 percent annually. This is a result of a national program for birth control suggested by the government for improving living standards.

CONCLUSION

Since Indonesia's energy reserves are large enough to cover the next two to three decades, the production of domestic energy will be sufficient to meet demand. Energy policy should be implemented with a view to the future limitations of nonrenewable resources. Adjustment of energy pricing should be implemented to maintain energy production activities, encourage conservation, and promote switching to more abundant fuels. For the foreseeable future, the role of energy in Indonesia will remain important in the balance of payments and the government budget.

PART V
Country Case Studies: South Asia

FOURTEEN

Energy and the Economy in Pakistan

YOUSAF KAMAL

Land, labor, and capital are the three traditional factors of production. Yet, the optimal utilization of these factors cannot be accomplished without energy in its various forms (oil, gas, coal, and electricity). The development of the energy sector is an essential component of social overhead facilities. The genesis of energy planning results from the desire for economic and social advancement of the people through balanced growth of the various sectors of the economy—industry, services, and agriculture. The availability of adequate energy resources is a sine qua non for production. Shortages of energy can have severe repercussions on national income, development of human and physical resources, and the overall pace of economic growth.

In Pakistan, the authorities dealing with energy at the federal level are the Ministry of Petroleum and Natural Resources and the Ministry of Water and Power. The executing agencies are the Oil and Gas Development Corporation, Pakistan Mineral Development Corporation, Pakistan Water and Power Development Authority, and Karachi Electric Supply Corporation, which work under the direction of their respective ministries. The Planning Commission is vested with the overall authority for providing policy guidelines for the exploration and development of energy resources based on requirements of oil, gas, coal, and electricity.

The scarcity of fossil fuel resources has been one of the major factors in Pakistan's economic development. As far back as 1981, Pakistan was producing 10 percent of its total oil requirements, with the remaining 90 percent imported at enormous cost in foreign exchange, also a scarce resource in Pakistan. In the past, energy supplies did not keep pace with energy demands, particularly for electricity and natural gas. The present shortages and load shedding have been the consequences. Oil production at present is meeting about 30–35 percent of total requirements. This may be viewed as a step forward, but it is still inadequate. The oil import bill has been continuously rising over the years and is likely to have a steep up trend in the future as oil-fired power plants are constructed to meet the ever increasing requirements for electricity. This will have a serious impact on Pakistan's already constrained balance-of-payments situation. Petroleum imports are estimated at 20 percent of export earnings, while the share of energy in total imports is estimated at about 13 percent.

Electricity production from hydroelectric plants has been increasing at about 10 percent, about the same rate of increase as in the installed generating capacity. In Pakistan, generation of electricity from hydroelectric plants is dependent upon both river flows, which fluctuate during the year, and water levels in reservoirs. The latter cannot be used for power production year-round because dams have been built primarily to supply water for irrigation needs. Power production is a secondary consideration. If all water were made available, hydroelectric generation would be far greater than that under the constraints discussed. However, the government endeavors to optimize production from hydroelectric power stations in order to minimize fuel cost from thermal power plants, as indicated by the comparatively lower growth rate of thermal generation (8.27 percent) against that of installed capacity (8.83 percent).

From 1980–1981 to 1985–1986, the growth rate of energy consumption was 7.8 percent; of GDP, 6.8 percent; and of population, 3 percent annually.

Electricity and LPG have recorded the greatest increase in consumption. The reported increase in electricity consumption (11 percent) is not real because of supply constraints resulting in load shedding. This is likely to continue until adequate thermal generation capacities based on oil have been constructed. The growth in LPG consumption has led to significant interfuel substitution in households that are using LPG instead of kerosene for cooking. Kerosene is also used for lighting in homes, particularly in rural areas that have not been supplied with electricity.

The increased electricity consumption is, to a substantial extent, responsible for increasing imports of petroleum products from 1.6 million tons of oil equivalent (toe) in 1980–1981 to 2.4 million toe in 1985–1986 (an annual growth rate of 8.4 percent). It may be worth mentioning that the unfavorable impact of this increase in imports of petroleum products on the balance of payments was partially mitigated by the declining prices of crude oil in the international markets.

Total commercial energy supply increased from 13.05 million toe in 1981 to 19.31 million toe in 1986 (an annual growth rate of 8 percent).

It is estimated that 84 percent of the urban and 30 percent of the rural population has been supplied with electricity. On an overall basis, about 45 percent of the country's total population has electricity in their homes. The per capita consumption of electricity was 195 kwh in 1986, rising from a bare 15 kwh in 1960 (an average annual growth rate of 7.6 percent).

It may be pertinent to give the sources of electricity production: oil (distillate furnace oil), high-speed diesel, natural gas, coal, and water. The major nonconventional source is nuclear power. The recoverable gas and coal resources are estimated at about 20.66 trillion cubic feet (ft^3) and 763.10 million tons, respectively, and the oil reserves at 233.76 million U.S. barrels. Until June 1985, 4.33 trillion ft^3 of gas had been consumed, leaving 16.33 trillion ft^3. The coal is not of good quality, and 90 percent of its consumption is by the brick kiln industry. The coal found at the country's major field, at Lakhra in Baluchistan Province, is lignite, which has a high sulfur content and other impurities, resulting in a low heating value of about 5,100 Btu/lb., half that of coal found in the United States, Canada, and other coal-producing countries. Studies are being carried out to use it for power production.

The hydroelectric potential of Pakistan is estimated at about 20,000 Mw. Of this, 10 percent has been developed at Tarbela, Mangla, and Warsak. Of the total 6,224 Mw generating capacity installed in the country in 1986, hydroelectric accounts for 2,897 Mw (47 percent) and thermal for 3,327 Mw (53 percent). The scarcity of fossil fuels for power generation has necessitated optimal exploitation of hydroelectric potential. A total of 1,928 Mw hydroelectric capacity is in various stages of installation at Tarbela (1,728 Mw) and Mangla (200 Mw).

It has also been planned to install 2,400 Mw at Kalabagh by 1996. A hydroelectric-ranking study carried out by foreign consultants indicated that 12,400 Mw of power can be obtained from development at Basha, Dasu, and Thakot (upstream of Tarbela Dam on the Indus River). In addition, it is planned to utilize the falls available at the barrages to produce about 420 Mw of cheap power. The development of hydroelectric power at Kalabagh and the barrages will save an estimated 3.36 million metric tons of furnace oil, at an annual savings of $395 million.

It is important to discuss the use of natural gas as fuel in power plants. In the early years of gas discovery and its use as fuel for power production, gas prices were low. The general thinking was that this source of electricity generation was abundant. Over time, it was realized that gas was underpriced compared with oil, which increased from $1.80 per barrel before October 1973 to $34 per barrel in October 1981. The price declined to $29 per barrel in March 1983. The January 1987 price was $16 per barrel. Therefore, the economic cost of gas was high. In addition, gas has a very high efficiency rate of about 80 percent when used in production of petrochemical products and fertilizers. The value added attributable to gas in production of chemicals and fertilizers was substantially higher compared with electricity production. Keeping these economic and other considerations in mind, a gradual policy of rationalizing gas rates has been pursued since July 1973. Gas prices have been increased from Rs 7.10 per 1,000 ft^3 in 1976 to Rs 38.54 per 1,000 ft^3 in 1986. Furthermore, since gas has a higher value-added contribution in the production of chemicals, synthetic fibers, and fertilizers, furnace oil will substitute for natural gas as fuel in the power plants. The sectoral mix of gas utilization has been changing rapidly; gas use in power generation has fallen, while its use in fertilizer production has risen.

In earlier years, the electricity shares of oil consumed by industrial and residential users resulted from interfuel substitution of gas for oil where gas was supplied at a preferential price. In the agricultural sector, interfuel substitution is between electricity and diesel oil for running irrigation tubewells. But this trend has been reversed by the government policy of adjusting gas prices to the world oil price. Another important policy change has been made in the cement industry, which previously used gas but has now been directed to use furnace oil. This was done to conserve gas for use in production where it has maximum energy-producing efficiency as well as value-added contribution.

It may also be appropriate to mention that indigenous gas resources have played a very significant role in Pakistan's energy supply picture in both the preshock and postshock periods, not only in supporting national fossil fuel requirements but also in alleviating the burden of foreign energy dependency. Gas energy and hydro resources have also been critical in achieving the policy goal of improved energy self-reliance.

Electricity consumption (sales) in 1986 by all categories of consumers (household, commercial, industrial, agricultural, and others) was 19.076 billion kilowatt-hours—38 percent for industries and the manufacturing sector; 15 percent for agricultural tubewells; 31 percent for households; 8 percent for shops and offices; and 8 percent for miscellaneous, including electric traction/railways. Overall, it is estimated that about 60 percent of electricity consumed went to the production of goods and services.

The share of households in electricity consumption has been continuously rising, from 26 percent in 1982 to 31 percent in 1986. This increase has resulted from higher living standards, reflecting greater use of electrical appliances and gadgets as per capita income has risen. Rapid urbanization and extension of electrical systems to new areas have also contributed to this shift. The decrease in agricultural consumption from 19 percent to 15 percent should not give the impression that agriculture, which is the mainstay of the national economy, has not fared well. On the contrary, it is indicative of more efficient groundwater resource use in optimizing agricultural production and of increasing awareness by farmers of the benefits of improved irrigation management practices.

For a developing country like Pakistan, energy prices have a direct bearing on the pace of development and on the socioeconomic conditions of the people at large. Energy, in its various forms, is used in manufacturing and farming. Tubewells for irrigation and reclamation of farmland, because of the twin menace of waterlogging and salinity, are operated by electricity. Diesel-run tubewells have been installed in areas where there are no electric facilities. More and more diesel-operated tubewells are being converted to electric power because of its convenience and low rates. Therefore, the requirements for electricity are increasing and power supply/demand deficits have been experienced. The policy of the government has been to accelerate the process of industrialization based on the latest technologies and to promote mechanization of farming. With these objectives, the government has been rather cautious in fixing energy and fuel prices. As a policy, the energy prices should be determined by their scarcity or opportunity cost. It is also important to allocate scarce capital investment resources for optimal utilization and maximum economic gains.

Keeping in view the broad national economic goal to accelerate development, the energy prices prior to the first oil shock were generally kept at fairly low levels, except for certain petroleum products (such as gasoline) that were subject to high indirect taxation. The other reason for higher gasoline prices was the fact that only the more economically prosperous people owned cars. In economic jargon, demand was inelastic with respect to changes in price. The consumer prices for natural gas were also fixed at levels significantly below those of petroleum products in order to provide adequate incentives for natural gas utilization in the industrial and power sectors. However, electricity pricing has historically been outside the sphere of general energy pricing. Although governed by similar social and political considerations, it has been largely based on the need to ensure an adequate financial return for the power utilities, which have been contributing 35 to 38 percent of their own financial resources (internal cash generation of funds from the sale of power) to the annual development (investment) program.

The level and structure of electricity tariffs have become distorted over time due to increased production costs. These higher costs are mainly a result of the increased share of the thermal component in the generation mix. The government has occasionally allowed tariff increases to enable the utilities to meet their financial obligations of self-financing the development and expansion program. However, the rates have not been specifically linked to long-run marginal cost (LRMC). Some reforms have been introduced in the tariff on domestic consumption. Block tariffs (the higher the electricity consumption, the higher the tariff rate) have been in force since 1976. This may be viewed as a step toward rationalizing and restructuring electricity rates. The policy of a comparatively low tariff for bulk consumers has been continued. This policy reflects the government's underlying belief that rapid growth in energy consumption is a necessary concomitant of industrial expansion and agricultural mechanization.

The policy of deliberately keeping energy prices low required review after the oil crisis. Modifications were made to take into account not only the higher price of imported oil but also the impact of higher international oil prices on the opportunity cost of major indigenous commercial fuels (natural gas and hydroelectric power). Accordingly, energy prices have been gradually raised toward the international level. As stated earlier, in the power sector, gradual adjustments in the tariff structure have been attempted to reflect LRMC. The government has protected certain key sectors—transport, industry, and low-income consumers—from the impacts of higher energy prices. Regulation and control of energy prices have focused on oil, gas, and electricity, leaving other energy resources (coal and firewood) more or less to the private market mechanism. Average energy prices in Pakistan increased at an annual rate of 5.8 percent in the preoil shock period (1965/1966–1972/1973), and by 18.9 percent in the postoil shock period (1973/1974–1980–1983). Oil prices increased more than gas prices because they were linked more directly to their import prices for balance-of-payments reasons. But the average prices of oil and gas increased at a far slower rate than the world market prices. The government, in order to promote incentives for increased investment by the producers in domestic oil and gas exploration, increased the wellhead prices for new domestic oil and gas discoveries.

The major source of nonconventional energy is nuclear power plants. The construction of a complex of nuclear plants for a country deficient in fossil fuels has become a necessity for economic development. Regarding other nonconventional renewable sources of energy, Pakistan has launched a systematic program to develop new renewable energy sources. The first biogas unit was installed in 1974. As of 1986, 4,159 plants have been installed.

Programs are being planned in collaboration and coordination with USAID for using solar energy for heating, cooking, electrification, and water pumping. It is planned to install 100 small windmills for water lifting between 1988 and 1990. The United Nations Development Program has assisted in setting up three windmills in windy areas of Baluchistan, Lower Sind, and Punjab.

The important policy measures that have been initiated for attaining self-reliance in energy resources are the following:

- An accelerated program of exploration and development of drilling for oil and gas, launched with foreign collaboration though higher wellhead prices
- Fiscal incentives to local entrepreneurs for undertaking exploration activities
- The Planning Commission's energy section has been restructured and enlarged. A separate section, ENERPLAN, for formulation of medium- and long-term energy plans, has been created. USAID has provided necessary assistance through its consultants
- A program of energy conservation is to be launched, for which a separate agency, ENERCON, under the Planning Commission, has been set up with collaboration and assistance provided by USAID. The key steps involved are a system of energy surveys and audits; continued analytical work; training of experts; institutionalized financing for conservation investment, demonstration, and dissemination of information; and the setting up of efficiency standards
- Improving the efficiency of power sector operation by reducing transmission and distribution power losses, improving thermal efficiency, and load management
- Active exploration and development of coal resources for power generation to minimize the import oil bill
- Coal mining has been declared an industry, and mine-mouth coal-bearing areas have been designated as special industrial areas, with coal mining at par with other industries. Tax holidays, concessions, and remissions available to other industries allowed for the coal mining industry
- Restructuring of electricity tariffs to facilitate demand management measures
- Exploitation and development of large hydroelectric potential
- Development of nonconventional sources of energy (biogas, solar, wind power).

FIFTEEN

The Energy-Economy Link in Sri Lanka

G. B. A. FERNANDO AND
D. A. U. DARANAGAMA

ENERGY SECTOR PERFORMANCE

The energy policymaking authority in Sri Lanka has many arms, because the harnessing of the different energy sources—petroleum, electricity, and fire—is vested in different organizations. Importation, refining, and distribution of petroleum products are vested in the Ceylon Petroleum Corporation. The distribution of LP gas is vested in the Colombo Gas and Water Company, Ltd., a business owned by the government of Sri Lanka. The investigation, development, and harnessing of hydroelectric and thermal sources of power generation are vested in the Ceylon Electricity Board. The development, exploitation, and conservation of forest resources are vested with the Conservator of Forests. The harvesting and marketing of timber resources are the functions of the State Timber Corporation. At present, there is no single authority in Sri Lanka for the importation and distribution of coal. The Atomic Energy Authority has a watching brief on nuclear power generation that will not be practically applied in Sri Lanka for at least two decades. Petroleum prices are determined by the Cabinet on the advice of the Ceylon Petroleum Corporation. Electricity prices are determined by the Cabinet on the advice of the Ceylon Electricity Board. Firewood prices are determined by market forces.

The main source of funds for the development of the energy sector is the public sector, supported by foreign assistance from friendly countries and international organizations. The availability of funds for energy development is determined in accordance with the priorities laid down by the Public Investments Program of the government. In the admission of new projects, the government has followed the two-stage approval procedure in which all projects have to be submitted to the Development Secretaries' Committee prior to submission to the

Cabinet. The government made a decision in 1984 that budgetary resources will not be allocated to activities that could be more easily and efficiently handled by the private sector. Operation and maintenance of existing assets to which adequate attention has, in the past, not been paid are now receiving due emphasis, and are provided for as a separate item under capital expenditure in budget estimates.

The importance of energy policymaking and its integration with economic planning and the policymaking process was realized soon after the first oil crisis of 1973, when the country suffered a severe impact on its balance-of-payments position. Immediate steps were taken to find solutions. In February 1974, the Ministry of Planning and Economic Affairs convened working groups to formulate an energy policy for Sri Lanka. This was followed in September 1975 by the establishment of the Development Planning Program and the Development Planning Unit in the Ministry of Planning and Economic Affairs with U.N. Development Program assistance. One of the priorities of the Development Planning Program was to establish the Energy Policy Planning Cell, which would undertake some of the basic studies needed in energy planning and development. With the assistance of Dr. P. D. Henderson, professor of political economy at University College, London, and Mr. T. L. Sankar, energy planner of the Planning Commission of India, invited through consultancies offered by the U.N. Development Program, basic studies on energy policy planning were completed and presented in November 1977 at a seminar on energy policy planning in Sri Lanka, which was attended by about 200 officials concerned with energy issues in the government as well as the private sector. The findings and recommendations were published in March 1978, followed by government initiation of several crucial steps toward energy sector development.

The APO Symposium on Energy Management held in Pattaya, Thailand, in August 1977 made concrete recommendations for the government and firm levels regarding new energy sources. Those recommendations were also incorporated into the above studies.

In 1979, the Cabinet appointed the Energy Study Committee to report on the short- and long-term energy requirements of the country, taking into account the recommendations made by the March 1978 U.N. Development Program study. Largely as a result of these recommendations, the Ministry of Power and Highways was reconstituted in February 1980 as the Ministry of Power and Energy. Highways became a separate ministry. The Natural Resources, Energy, and Science Authority was established by an act of Parliament in June 1982.

The National Resources, Energy, and Science Authority undertakes research and development activities on energy. The Ministry of Power and Energy supervises the Ceylon Electricity Board; Ceylon Petroleum Corporation; and Colombo Gas and Water Company, Ltd. It promotes energy conservation and is also the focal point for new and renewable sources of energy. An energy conservation fund was established in 1985. Following the U.N. conference on new and renewable sources of energy in August 1981 and the ESCAP regional follow-up in March 1982 in Colombo, a national consultative committee on new and renewable sources of energy was covened in 1983 by the Ministry of Power and Energy in order to identify projects that would require external financial assistance, prepare project proposals, collect information, and identify institutional arrangements for project implementation. Some projects have already been recommended and funded.

The Ceylon Electricity Board intensified its efforts to develop the demand and supply positions for electricity generation, transmission, and distribution. The Ceylon Petroleum Corporation actively monitored the supply and demand positions for petroleum products in Sri Lanka and optimized the output from the refinery to meet national needs. The United Nations Environment Program sponsored a Rural Energy Demonstration Center in 1975 at Pattiyapola, Sri Lanka, to utilize renewable sources of energy to meet the basic needs of a typical rural village. The program is executed by the Ceylon Electricity Board.

Until recently, the main interests in the forestry sector were conservation and law enforcement. It was only in 1978 that the sector's importance in protecting the environment and in supplying domestic energy and industrial timber needs was realized. As a result, government investment in the sector increased from about Rs 15 million in 1978 to about Rs 220 million in 1986. In recognition of the fact that the forest cover was rapidly being depleted, several reforestation programs were initiated after 1980 with foreign assistance. Between 1977 and 1986, about 64,000 hectares were reforested. Another 16,000 hectares are likely to be reforested between 1987 and 1991.

A five-year implementation program is being prepared under a draft Forestry Master Plan. Six projects have thus far been identified under the plan:

- Industrial and protection plantations
- Forestry extension
- Farmer woodlots
- Block firewood plantations
- Forestry protection and management
- Forest education

Government investment under the plan would run about Rs 225 million annually from 1986 to 2000. It is expected to generate substantial employment opportunities and a net foreign exchange savings through increased production of industrial timber and firewood, as well as through prevention of losses in the production process. The total effect would be RS 1 billion in 1990, increasing to Rs 2.5 billion in 2000.

The decision to accelerate the implementation of the Mahaweli Development Program was taken in order to find early solutions to three major problems—high unemployment, food shortages, and energy shortfalls—faced by the country in 1977. Under the accelerated Mahaweli Program, it was envisaged to provide irrigation for about 120,000 hectares of new land and to add about 500 Mw of installed hydro power capacity to the national grid. It was also thought that about 120,000 farm families could be settled on the new lands.

Oil prices, which fell from about U.S.$30 per barrel in the early 1980s to around U.S.$13 per barrel in 1986, are not expected to remain at that level for long. The price is expected to gradually rise. As the value of the U.S. dollar with respect to the Sri Lankan rupee continues to rise, the relative reduction in oil prices decreases in terms of the Sri Lankan rupee. The local prices of petroleum products, except furnace oil used in industry, remain unchanged because low prices are considered only a temporary phenomenon. The price of furnace oil used in industry has been reduced by 20 percent. The prices of petroleum products other than furnace oil have not changed since July 22, 1983.

Any profits are absorbed by the government through taxation of petroleum products. Maintenance of petroleum prices at current levels will not induce increased demand for petroleum products, which is in keeping with the policy of reducing dependence on imported oil products. Petroleum prices are fixed so that total sales of petroleum products will just cover import costs and operational expenses, including taxes. There are no known resources of petroleum in Sri Lanka, although investigations have been conducted for oil. Petroleum supplies are maintained in accordance with storage capacity.

Lower oil prices have not brought about any major changes in energy administration or policy. The main concern of the government has been to reduce the oil import bill, which reached very high proportions in 1982.

It is significant to note that the percentage of net oil imports to nonpetroleum exports:

- Remained around 4–5 percent before the 1973 oil crisis
- Rose to around 15 percent between the two oil crises of 1973 and 1978
- Rose to about 35 percent between 1979 and the oil price decrease in 1985, reaching a peak of 47 percent in 1982
- Declined to 9 percent in 1986, when oil prices decreased.

A significant feature is that the volume of petroleum imports remained at around the same level; only the prices have fluctuated. Therefore, it is assumed that the volume of petroleum imports will remain at around the same level if prices are not changed. Prices have therefore not been changed in order to check the demand for petroleum products. Another feature is that the impact of the oil prices on the balance-of-payments position is similar to that of the pre–1973 era.

With the sudden increase in the price of oil in 1973, the government took steps to reduce its dependence on petroleum imports. The main features have been the following:

- To identify the supply-demand balance of energy sources. It was only in 1978 that firewood was recognized as the major supplier of energy—meeting about 70 percent of demand—and as being rather inefficiently used in the domestic sector.

- To harness available hydroelectric resources. The government embarked on the Accelerated Mahaweli Ganga Development Program, and continues to harness other hydro resources, such as the Samanalawewa Project. It also is studying the hydroelectric resources remaining.
- The government intensified its investment in the forestry sector, from Rs 15 million in 1978 to Rs 220 million in 1986, through several reforestation programs and developed the Forestry Master Plan to arrest further depletion of forest cover, to protect the environment, and to supply domestic energy and industrial timber needs.
- Thermal power plants, needed to back up hydroelectric power plants, use residual fuel oil from the oil refinery, thus maximizing utilization of oil imports.
- Petroleum prices have been revised to achieve the goals of pricing kerosene at affordable prices; pricing diesel as low as possible, because it feeds the mass transport system; fixing gasoline prices to offset petroleum subsidies; and covering import costs and operational expenses, including taxes.
- Electricity prices have been regularly revised in order to meet an annual 8 percent rate of return on revalued assets, provide relief to small electricity consumers in the domestic sector, subsidize rural electrification schemes, and structure tariffs to industrial consumers based on energy consumption and maximum demand.
- Projects on energy conservation and fuel substitution, particularly in the industrial sector, have been promoted with assistance from the United Nations, USAID, Swedish IDA, CHOGRM (the Netherlands), and the creation of an energy conservation fund.
- Activities on new and renewable sources of energy have been centralized and promoted. A national consultative committee identifies need-based and impact-producing projects requiring external financial assistance, prepares project proposals, collects information, and identifies institutional arrangements for modalities of project implementation.
- The Ministry of Power and Energy and the Natural Resources, Energy, and Science Authority have been created in order to centralize and institutionalize activities in energy coordination, planning, and development, keeping in mind the impact on the environment.
- The need to increase the efficiency of petroleum product use in the transport sector by easing traffic congestion, improving road maintenance and traffic flow within cities, and maintaining dual lanes in intercity highways is seen for the short and medium terms. The Road Development Authority has been established, and many programs are under way. The Ministry of Transport has embarked on studies in this area with World Bank assistance, and a report is imminent.
- The impact of energy development on the environment is under independent study by the Central Environment Authority. The National Conservation Strategy has completed its task, and a report is expected to be released shortly.
- A major rural electrification program has been embarked on by the government. The total number of village electrification schemes increased from 943 at the end of 1977 to 5,946 at the end of 1986.

The main purpose of these activities is to reduce dependence on foreign energy sources and optimize locally available energy sources. There is no perception of a change in these activities under lower oil prices.

Coal-fired electricity generation plants have been targeted for commissioning in 1993. If oil prices continue to remain low during the 1990s, the feasibility of converting the coal-fired electricity generating plants to dual-fired plants that can run on coal as well as furnace oil may be examined to take advantage of the reduction in oil prices. The electricity generation expansion plan is likely to be affected in the longer term. Chemical industries based on oil refinery outputs such as urea are to be reexamined. Export-oriented industries could alter their plans. Plans on energy conservation are unlikely to be affected. Table 15.1 provides estimates of energy consumption over the period 1986–1996.

Noncommercial sources of energy will play an important role in the near future. Efficiency in the use of these sources, particularly in the domestic sector, which utilizes about five-sixths of the available noncommercial resources, is in the very low region of 7 percent. There is considerable scope in increasing the efficient use of this resource. With the increase in affluence, the demand for noncommercial sources decreases. However, the total number of households in the rural sector increases with population and, for this reason, the demand for noncommercial sources will increase. The overall effect is that the useful energy derived from noncommercial energy sources will increase while the share of noncommercial energy in total energy use will decrease.

EVIDENCE OF STRUCTURAL CHANGES IN ENERGY AND ENERGY/ECONOMY RELATIONSHIPS

Soon after the oil price increase in 1973, the Ministry of Planning and Economic Affairs set up working groups to formulate an energy policy for Sri Lanka. Saturday was declared a public holiday and a five-day workweek was introduced in order to save transport fuel.

Energy policy planning was given priority in the Development Planning Program set up in the Ministry of Planning and Economic Affairs in September 1975 with U.N. Development Program assistance. Until the 1973 oil crisis, the Ceylon Electricity Board and Ceylon Petroleum Corporation determined their programs independently. At times, their pricing policies were based on different and sometimes contrary goals. Only when energy policy planning was given priority in the Ministry of Planning was it realized that the activities of the Ceylon Electricity Board and Ceylon Petroleum Corporation should be coordinated in order to achieve common energy goals. And it was only in 1978 that the significance of firewood as a major source of energy was understood. The need to coordinate and centralize energy issues through the establishment of an energy ministry or an energy authority soon gained acceptance.

Key recommendations that the Energy Policy Planning Unit in the Ministry of Planning and Economic Affairs made in March 1978 were the following:

Table 15.1
Energy Consumption Estimates, 1986–96

	1986	1987	1988	1989	1990	1991	1992	1993	1994	1995	1996
Petroleum import (KMT)	1,648	1,552	1,634	1,700	1,906	1,907	—	—	—	—	—
Value (SDR millions)	161	171	192	218	246	258	—	—	—	—	—
Natural gas	—	—	—	—	—	—	—	—	—	—	—
Coal (KMT)	30	30	30	30	30	30	30	390	390	750	750
Hydro (Gwh)	2,671	2,980	3,080	3,288	3,383	3,805	3,805	3,805	3,805	3,805	3,805
Nuclear	—	—	—	—	—	—	—	—	—	—	—
Noncommercial energy (toe)	4,158	4,216	4,275	4,334	4,493	4,452	4,498	4,544	4,590	4,636	4,683

KMT = Thousand metric tons.

Gwh = Gigawatt-hours.

toe = Tons of oil equivalent.

Source: Ministry of Power and Energy, Government of Sri Lanka, Colombo, 1987.

- That the pattern of petroleum demand and supply from the refinery was mismatched, and a pricing policy with respect to electricity as well as oil products needed to be studied. In 1979, the Committee of Development Secretaries appointed the Fuels Committee, which made this study and provided a list of recommendations.
- That thermal sources of power generation were needed to avoid power shortages in 1980 and 1981.
- That energy conservation measures need to be begun as early as possible at the government and firm levels.
- That hydroelectric resources should be developed to meet the energy needs of the country. There is no "excess" electric power to be sold to India, as was once believed in some engineering and official circles.
- That priority should be given to the development of forest resources in order to arrest further denudation of the forests and prevent desertification.
- That the development of alternative energy sources should be encouraged and monitored.
- That it was possible for Sri Lanka to organize these studies with a minimum of external assistance.

Government allocation to the forestry sector increased greatly. Many activities of the forestry sector were embarked on with external assistance. (These are described elsewhere in this chapter.) The government embarked on the Accelerated Mahaweli Ganga Program, which is handled by a separate ministry in the Ministry of Mahaweli Development. The Mahaweli Development Board and the Mahaweli Economic Agency were instituted under the Ministry of Mahaweli Development. The Forestry Resources Development Project was set up in the Ministry of Lands and Land Development to formulate a Forestry Master Plan. The Ceylon Electricity Board, Ceylon Petroleum Corporation, Colombo Gas and Water Company, and Energy Conservation Fund function under the purview of the Ministry of Power and Energy, which was established in 1980. The Ministry of Power and Energy is also the focal point for new and renewable sources of energy, as well as for activities connected with the Economic and Social Commission for Asia and the Pacific's (ESCAP) U.N.-funded Regional Energy Development Program. The ministry also coordinates activities on energy conservation. The Coal Power Development Project was instituted in the Ceylon Electricity Board. The Ministry of Transport embarked on a study on transport policy with assistance from the World Bank.

The oil refinery planned to expand capacity in its processing facilities from 1.78 million metric tons of crude oil per year to 2.2 million metric tons in 1973. Because demand decreased with the oil price increases in 1973, the expansion of the refinery was effected only in 1979. A singe-point mooring buoy was installed by the Ceylon Petroleum Corporation for direct discharge of crude oil from supertankers.

Although petroleum prices have increased more than tenfold since 1972, the volume of petroleum sales has not changed much. Kerosene sales have decreased with the fourfold increase in price in 1980 and the further twofold increase in price in 1982. Although prices of auto diesel have increased nearly twentyfold since 1972, auto diesel sales have been steadily increasing because that fuel is used in mass transport.

It was realized in 1978 that if energy conservation measures were introduced at the government and firm levels, a reduction in petroleum imports could be achieved. A World Bank mission to Sri Lanka in 1981 reiterated that a Rs 200–300 million program of energy conservation could bring annual savings in the order of Rs 300–500 million, and a successful program could confine the oil import bill in 1985 to around 45 percent of export earnings. Based on the World Bank recommendation, the United Nations Development Program supported the Industrial Energy Conservation Program, which trained energy managers in leading industrial institutions and conducted energy audits. No- and low-cost energy conservation measures were executed. However, conservation programs that required large investments could not be begun because of financial constraints. The Ceylon Petroleum Corporation has achieved substantial savings by introducing energy conservation measures. Although energy conservation targets were not fully achieved, the oil import bill fell to 18 percent of export earnings in 1985, mainly because of declining international oil prices. If conservation measures are properly introduced, the cost of oil imports could be reduced even further. The impact of the oil bill on the balance-of-payments position has now reached pre–1973 levels.

The government policies and regulations that have induced structural changes are summarized below:

- Gasoline, kerosene, and diesel prices were increased periodically.
- Kerosene stamps were introduced to provide relief to poor consumers.
- The license fee on diesel vehicles was increased to discourage the use of private diesel vehicles.
- Electricity prices were restructured and a fuel adjustment charge was introduced.
- The import duty on motorcycles was reduced following the recommendations of the 1980 fuel committee report.
- Private vehicle use on Sundays between 8:00 A.M. and 2:00 P.M. was banned. However, the ban was lifted after a few months because gasoline sales decreased and diesel sales increased, thereby affecting the cross-subsidy between gasoline sales and cheap kerosene for poor consumers.

Energy conservation assistance has been provided by various foreign agencies. Energy audits have been conducted at major firms producing tires, textiles, steel, paper, plywood, rubber, tea, chemicals, flour, cement, ceramics, hardware, and sugar. Energy managers have been trained in Sri Lanka as well as at specialized institutions in developed countries. The energy managers have recommended energy conservation programs in their institutions. Low- and no-cost energy conservation programs have been implemented. Energy conservation programs that require large investments will be implemented when funds become available.

Firewood-efficient stoves are being promoted in the domestic sector throughout the country.

CHANGING OIL PRICES AND ECONOMIC DEVELOPMENT

Electricity and petroleum prices needed radical restructuring after the oil price increase in 1973 to balance investments and revenue in the energy sector. As a result, these prices underwent frequent revisions in order to generate sufficient revenue to meet investment needs as well as to provide cross-subsidies to poorer consumers needing energy supplies to meet basic needs.

Petroleum prices have been revised to achieve the goals of affordable pricing of kerosene, largely used for lighting in rural areas, pricing diesel as low as possible because it fuels the mass transport system, fixing gasoline prices to offset petroleum subsidies, and covering import costs and operational expenses, including taxes.

Electricity and petroleum prices stabilized in 1985 in order to just cover the cost of sales (including taxes) and to meet many other criteria, such as providing relief to poor consumers. However, the pricing has not covered certain facilities provided for energy consumers, such as maintenance of roads and highways. If such costs are also included, the present prices will have to be increased.

The decrease in oil prices in 1986 is considered only a temporary phenomenon. It is difficult to predict how long prices will remain at these levels. At these prices, net oil imports cost around 9 percent of the export earnings of the country. One objective the government decided on after the oil price increase in 1973 was to reduce oil import dependence. If energy tariffs are lowered, the demand for petroleum products will increase and it will be difficult to curtail demand when oil prices increase once again. Besides, energy tariffs do not cover such additional expenditures as road maintenance. Under these circumstances, energy prices have not been lowered. The exception is furnace oil for industrial use, which has been reduced by 20 percent. When oil prices were high, the government had to curtail many development programs because of heavy oil bills. When oil bills are low, the government can use the funds according to its priorities.

The decline in oil prices has not generated large amounts of revenues for Sri Lanka. On the other hand, it has reduced oil import bills; therefore the government is in a position to embark on those programs it previously had to curtail in order to meet high oil import bills.

Trade patterns with oil-exporting countries have not been seriously affected. However, employment opportunities for Sri Lankans in oil-exporting countries have been reduced.

CURRENT ECONOMIC DEVELOPMENT PLANNING AND STRATEGIES

In this section, the important government polices and strategies that may affect future economic development are summarized.

Energy conservation, power generation planning, and the search for alternative sources of energy are priority considerations in the energy sector. An increase in the supply of firewood through afforestation and popularization of modern and efficient techniques of firewood utilization are given high consideration. The consumption of petroleum in the transport sector has been rapidly increasing in recent years. Therefore, more efficient modes of mass transportation will be needed in order to reduce dependence on foreign oil imports. A major study on transport policy has been completed with World Bank assistance. Coal-fired electricity generating plants are expected to be commissioned from 1993, while nuclear power plants are not expected to be commissioned before 2010. The pace of rural electrification has increased very rapidly since 1977. At present, it is estimated that only 24 percent of the households in Sri Lanka have access to electricity; therefore, the pace of rural electrification is expected to continue at a high rate.

Although firewood and agricultural residues comprise almost 70 percent of energy consumption in Sri Lanka, this source is used very inefficiently in the domestic sector. Eighty percent of consumed firewood is obtained from the nonforest sector. The forest cover has been halved since the 1960s. A shortage of firewood is expected in the dry zone by 1995 and throughout the country by 2010. In order to arrest the degradation of the forest cover, meaningful steps have to be taken very quickly with respect to land use planning and watershed management. Technologies in the use of agricultural residues have to be improved. Areas of interest in new and renewable sources of energy lie in solar photovoltaics, solar heating devices, wind energy for lift irrigation, biomass, and biogas, firewood and charcoal, and the rehabilitation of minihydro plants. A wide gap remains between the experimental and application stages in the development of alternative energy. The thrust has to be placed on the improvement, adaptation, application, standardization, and commercialization of mature technologies.

The development of nuclear power plants in Sri Lanka will be in the hands of those just entering the universities today. Scientists trained in nuclear power technology should be of the highest caliber available. It is necessary to train the best students in the sciences so that they may look forward to careers in the application of modern science to development programs, of which nuclear power programs will be only one. This calls for wise science policies with long-term commitment, generous patronage, self-governance, and free international contacts. Training should occur at the undergraduate as well as the graduate level.

The need to develop new and renewable sources of energy, as well as to promote energy conservation in order to reduce the dependence on foreign oil imports, has clearly been identified. The issue is whether the steps to achieve this are being taken in an orderly, peaceful, progressive, just, and integrated manner. This transition entails the rationalization of the energy development process. It will also require significant and continuous progress in the efficient and rational use of available resources.

Bibliography

Ahn, B. H., H. W. Kim, D. M. Nesbitt, and R. L. Phillips. 1986. ''Integrated National Energy Planning: A Case Study of the Republic of Korea.'' *Energy Journal* 7:2, pp. 13–36.

Akamatsu, K. 1962. ''A Historical Pattern of Economic Growth in Developing Economies.'' *The Developing Economies* 1, preliminary issue no. 1, pp. 3–25.

Alchain A., and S. Woodward. 1987. ''Review of O. E. Williamson's 'The Economic Institutions of Capitalism'.'' Memo.

Alejo, Leander. 1983. *A Study of Energy-Economy Interaction in the Philippines*. Monograph Series no. 3. Manila: Philippine Institute for Development Studies.

———. 1986. *Energy Demand Management Policies in the Philippines*. Energy Study Series no. 9. Manila: Asian Development Bank.

Andersson, T., and B. Ternstrom. 1987. ''Private Foreign Investments and Welfare Effects: A Comparative Study of Five Countries in Southeast Asia.'' Stockholm: Stockholm School of Economics Research.

Arismunadar, A. 1986. ''Recent Energy Developments in Indonesia.'' Paper delivered at the Asian Forum on Energy Policy. Bangkok, October 27–30.

Asian Development Bank. 1987. *Energy Policy Experiences of Asian Countries*. Manila: The Bank.

Asian Oil and Gas. ''Malaysian New Terms Cap ASCOPE '85.'' January 1986, pp. 6–8.

Balassa, B. 1977. ''A 'Stages Approach' to Comparative Advantage.'' World Bank Staff Working Paper no. 256. Washington, D.C.: World Bank, May.

Barrows, G. H. 1983. *Worldwide Concession Contracts and Petroleum Legislation*. Tulsa, Okla.: Pennwell Publishing Company.

Bautista, R. M., J. H. Power, and Associates. 1979. *Industrial Promotion Policies in the Philippines*. Manila: Philippine Institute for Development Studies.

Berndt, E. R., and D. O. Wood. 1975. ''Technology, Prices, and the Derived Demand for Energy.'' *Review of Economics and Statistics* 57: 259–268.

———. 1979. "Engineering and Econometric Interpretation of Energy-Capital Complementarity." *American Economic Review* 69: 342–354.

Blitzer, C. R., P. E. Cavoulacos, D. R. Lessard, and J. L. Paddock. 1983. "Overcoming the Oil and Gas Exploration in Developing Countries." MIT Energy Laboratory Working Paper no. 500225. Cambridge, Mass.: The Laboratory.

———. 1985. "Oil Exploration in the Developing Countries: Poor Geology or Poor Contracts?" *Natural Resources Forum* 9, no. 4 (November), pp. 293–302.

Briggs, C. K., and I. Y. Borg. 1986. *U.S. Energy—1985*. Report UCID–19227–85. Lawrence Livermore National Laboratory. Livermore, Calif.

Campbell, B. 1982. "Petrodollar Recycling 1973–80, Part I: Regional Adjustments and the World Economy." Asian Development Bank Economic Staff Paper no. 8. Manila: The Bank.

Chamber, M. M., and W. S. Chern. 1985. "An Energy-Economy Model for Pakistan." Paper presented at the Seventh Annual North American Meeting of the International Association of Energy Economists, Philadelphia.

Chern, W. S. 1987. "Aggregate Energy and Oil Demand in Less Developed Countries." *Energy Systems and Policy* 10:73–91.

Chern, W. S., and H. Soberon-Ferrer. 1987. "Structural Changes of Energy Demand in Less Developed Countries." *OPEC Review* 11:71–86.

Cheung, S. N. 1969. *The Theory of Share Tenancy*. Chicago: University of Chicago Press.

Choe, B. J. 1978. "Energy Demand Prospects in Non-OPEC Developing Countries." In *Workshops on Energy Supply and Demand*. Paris: International Energy Agency.

———. 1980. "Energy Demand in Developing Countries." In Joy Dunkerly, ed. *International Energy Strategies, Proceedings of the 1975 IAEE/RFF Conference*. Oelgeschlager, Germany: Grunn & Haim.

Coase, R. H. 1952. "The Nature of the Firm." In George Stigler and Kenneth Bounding, eds., *Readings in Price Theory*. Homewood, Ill: Richard D. Irwin. (First published 1937.)

———. 1960. "The Problem of Social Cost." *Journal of Law and Economics* 3 (October), pp. 1–44.

Ebashi, M., and M. Hishida. 1986. "China's Open Door Policy and Its Implications for Asia-Pacific Countries Towards the Year 2000." Paper presented at Beijing Conference on the Asia-Pacific Economy Towards the Year 2000, Kuala Lumpur, November.

Economist Intelligence Unit. 1986. *1986/87 Yearbook: Energy for East Asia and Australia*. London: The Economist.

Energy Economics Research, Ltd. May 1987. *Oil and Energy Trends: Annual Statistical Review* 1970–1986. *Statistical Review 1987*. Reading, Great Britain.

Energy Information Administration, U.S. Department of Energy. 1982. *Energy Company Development Patterns in the Postembargo Era*, DOE/EIA–0349/1&2. Washington, D.C.: Resources for the Future.

———. Various Years. *Performance Profiles of Major Energy Producers*, DOE/EIA 0206.

Fesharaki, F., and D. Isaak. 1986. "The World Oil Market: A Brief Note." Honolulu: Resource Systems Institute, East-West Center.

———. 1986. "OPEC and the World Refining Crisis." EIU Special Report no. 168. London: Economist Intelligence Unit.

Fesharaki, F., and H. Razavi. 1986. "Spot Oil, Netbacks, and Petroleum Futures." EIU Special Report no. 1063. London: Economist Intelligence Unit.

Fesharaki, F., D. Isaak, S. Pintz, and N. Yamaguchi. 1986. "Pacific Oil Study: Phase One (SPOS I)." Honolulu: Energy Program, East-West Center.

Fuller, W. A., and G. E. Battese. 1974. "Estimation of Linear Models with Cross-Error Structure." *Journal of Econometrics* 2:67–78.

Government of Malaysia. 1982. "An Introduction to Malaysia's Energy Sector." Kuala Lumpur: Ministry of Energy, Telecommunications, and Posts.

———. 1985. "National Energy Balance, Malaysia 1978–1984." Kuala Lumpur: Ministry of Energy, Telecommunications, and Posts.

———. 1986. "Fifth Malaysia Plan, 1986–1990." Kuala Lumpur, Malaysia.

Government of Malaysia/UNIDO. 1985. "Medium- and Long-term Industrial Master Plan, 1986–1995." Vol. II, pt. 5, "Chemical and Petrochemical Industry."

Griffin, J. M., and P. R. Gregory. 1976. "An Intercountry Translog Model of Energy Substitution Responses." *American Economic Review* 66:845–857.

Hsiao, F. S. T., and M. W. Hsiao. 1985. "Elasticities, Ratios, and Energy Modelling." *Energy Economics* (July), pp. 153–58.

Hudson, E. A., and D. W. Jorgenson. 1974. "U.S. Energy Policy and Economic Growth, 1975–2000." *Bell Journal of Economics* 5:461–514.

Institute for Energy Analysis. 1979. *Economic and Environmental Impacts of a U.S. Nuclear Moratorium, 1985–2010*. Cambridge, Mass.: MIT Press.

Iwayemni, A. 1987. "Investment in Oil Exploration and Development in Asia: The Implications of Recent Energy Market Developments." *OPEC Review* 11, no. 1 (Spring), pp. 87–104.

James, W., and R. Young. 1987. "China's Modernization Program and Its Likely Effects on the Asia-Pacific Regional Economy." Resource Systems Institute Working Paper 87–1. Honolulu: East-West Center.

Kim, H. T. 1986. "Energy Demand Management Policies in Korea." Energy Study Series no. 1. Manila: Asian Development Bank.

Korea Energy Economics Institute. 1987. "Energy Future for Korea: Long-Term Outlook and Strategies, 1989–2010." Seoul: The Institute.

Maddala, G. S. 1971. "The Use of Variance Components Models in Pooling Cross Section and Time Series Data." *Econometrica* 39:341–358.

Mikesell, Raymond F. 1984. *Petroleum Company Operation and Agreements in Developing Countries*. Baltimore: Johns Hopkins University Press.

Ministry of Energy. 1980. *Ten-Year Energy Program, 1980–1989*. Manila: The Ministry.

———. 1985. *National Energy Program, 1986–1990*. Manila: The Ministry.

Mohd Ali, Rozali B. 1986. "Energy Cooperation in ASEAN: Status and Outlook." Paper presented at Conference on Regional Development and Security: The Ties That Bind. Second Meeting of ASEAN Institutes of Strategic Studies, January 12–16.

National Economic and Development Authority. 1986. *Medium-Term Philippine Development Plan, 1987–1992*. Manila: The Authority.

Naya, S. 1987. "Trade and Trade Policies." In S. Ichimura, ed., *Development Strategies and Productivity Issues in Asia*. Tokyo: Asian Productivity Center.

Naya, S., and E. Ramstetter. 1987. "United States Direct Foreign Investment in Asia's Developing Economies." Paper prepared for the Conference on Economic Cooperation Through Foreign Investment Among Asian and Pacific Countries, Bangkok, May.

Pindyck, R. S. 1979. *The Structure of World Energy Demand*. Cambridge, Mass.: MIT Press.

Ramstetter, E. 1986. "The Impacts of Direct Foreign Investment on Host Country Trade and Output: A Study of Japanese and United States Investment in Korea, Taiwan, and Thailand." Ph.D. dissertation, University of Colorado, Boulder.

Roumasset, James A. 1988. "Agency Theory, Agricultural Firms, and the New Institutional Economics." Department of Economics, University of Hawaii, Working Paper no. 1. Honolulu: The Department.

Song, C. Y. 1987. "Estimation of Capital Stocks Series in Korea." Master's thesis, Department of International Economics, Seoul National University.

Strout, A. M. 1983. "Estimating Energy Consumption from Cross Country Relationships." Department of Urban Studies and Planning, Massachusetts Institute of Technology.

Suryadi. 1985. "Energy Consumption Pattern in Indonesia." Paper presented to seminar on rational use of energy, São Paulo, Brazil, November 24–30.

Wallace T. D., and A. Hussain. 1969. "The Use of Error Components Models in Combining Cross Section with Times Series Data." *Econometrica* 37:55–68.

Wallis, J. J., and D. C. North. 1987. "A Framework for Integrating Transaction Costs and Institutional Change into Economic History." Department of Economics, University of Maryland, College Park.

Wijarso, and Adin Adimir. 1987. "General Analysis of Crude Price: Pre–21st Century." SEN III/1987 KNI-WEC. Jakarta: BAPPENAS.

Williamson, O. E. 1985. *The Economic Institutions of Capitalism: Firms, Markets, and Relational Contracting*. New York: Free Press.

World Bank. 1982. *Philippine Energy Sector Survey*. East Asia and Pacific Regional Office Projects Department, in association with the Asian Development Bank. Manila: World Bank.

———. 1987. "Indonesia Energy Option Review." Jakarta: World Bank.

Index

Contributors

WEN S. CHERN is Professor, Department of Agricultural Economics and Rural Sociology, Ohio State University. He is the author of numerous articles and books on energy demand and the interactions between economic growth, industrialization, and the energy sector.

D. A. U. DARANAGAMA is Energy Conservation Officer, Ministry of Power and Energy, Government of Sri Lanka. He holds a diploma in environmental engineering from the University of Moratuna and a Bachelor of Science degree in engineering from the University of Peradeniya, Sri Lanka.

GABRIEL B. A. FERNANDO is Director, Energy Planning, Ministry of Power and Energy, Government of Sri Lanka. He holds an M.A. degree in mathematical physics from the University of Cambridge. He has served as a consultant to the Asian Development Bank, the United Nations Development Program, and the United Nations Economic and Social Commission for Asia and the Pacific.

FEREIDUN FESHARAKI is Research Associate, Resource Systems Institute (RSI), East-West Center and Leader of RSI's Energy Program. He has published extensively on the petroleum industry and has served as advisor to numerous governments and energy companies in the Asia-Pacific region.

SHARON HOFFMAN is a Project Fellow with the Energy Program at the East-West Center in Honolulu, where she supervises the energy studies of the Philippines, Australia, and the Pacific Islands. Formerly she was a Research Fellow at Harvard University and has also been employed as a consultant to a number of Asia-Pacific and Middle Eastern governments, and oil companies. She holds an M.S. degree in public policy from Carnegie Mellon University.

WILLIAM E. JAMES is Research Associate, Resource Systems Institute, East-West Center, and Associate Professor, Department of Economics, University of Hawaii. He has published extensively on natural resources, trade, and development issues in the Asia-Pacific region and is a former staff member of the Asian Development Bank.

YOUSAF KAMAL is Director of Industries and Mineral Development, Government of the Punjab, Pakistan. He holds an M.P.A. degree from Harvard University and an M.S. degree in economics from the London School of Economics.

TORU KIMURA is an Economist at the Institute for Energy Economics, Tokyo. He has prepared several studies on energy supply and demand in the Asia-Pacific region and Japan.

BUNCHIRD KULPHANICH is Head, Energy Master Plan Section, National Energy Administration, Government of Thailand. She holds a bachelor's degree in accounting from Chulalongkorn University.

CAMILO JOSE LIM, JR., is an Economic Development Specialist with the National Economic and Development Authority in Manila. He holds an M.A. degree in economics from Williams College, Massachusetts. He has participated in preparation of numerous studies on energy pricing and planning in the Philippines.

WIROTE MANOPIMOKE is a Ph.D. candidate in the Department of Agricultural and Resource Economics, University of Hawaii. He has also been a research intern with the Resource Systems Institute, East-West Center.

SAHATHAVAN MEYANATHAN is Professor, Faculty of Economics and Administration, University of Malaysia and is associated with the Institute for Strategic and International Studies, Kuala Lumpur. He has published numerous articles on energy policy and economic development in Malaysia.

R. K. PACHAURI is Director of Tata Energy Research Institute in New Delhi. Dr. Pachauri has served as President of the International Energy Economics Association and has been a consultant to the Asian Development Bank. He has published extensively on energy issues.

HENDRO PRAWOTO is a senior member of the Oil and Gas Research Development Centre of Indonesia, where he heads the staff of the Planning and

Economics Department. He has extensive international experience in the economics of the petroleum industry and holds an M.S. degree from Azerbaijan Institute of Oil and Chemistry, USSR.

GOVIND P. SAHA is a senior consultant with Arthur Young Associates. He is a former Director of Electricity Industry, Ministry of Energy, Government of New Zealand. He has written numerous articles and reports on the economics of energy planning and conservation.

JEONG-SHIK SHIN is a fellow at the Korea Energy Economics Institute. He holds a Ph.D. in economics from the Ohio State University. He has published several articles on energy pricing and currently is involved in the economics forecasting of energy demand.

TSENG TSENG-TSAI is Project Planner and Controller, Energy Committee, Ministry of Economic Affairs, Republic of China. He holds an M.A. degree in economics from National Ching Hsing University.

www.ingramcontent.com/pod-product-compliance
Ingram Content Group UK Ltd.
Pitfield, Milton Keynes, MK11 3LW, UK
UKHW020815290926
15470UKWH00003B/1164

9 780275 931049